AF294946

Tucholsky Wagner Zola Scott Sydow Freud Schlegel
Turgenev Fonatne Wallace Walther von der Vogelweide Fouqué Friedrich II. von Preußen
Twain Weber Freiligrath Frey
Fechner Weiße Rose von Fallersleben Kant Ernst Frommel
Fichte Hölderlin Richthofen
Engels Fielding Eichendorff Tacitus Dumas
Fehrs Faber Flaubert Eliasberg Ebner Eschenbach
Maximilian I. von Habsburg Fock Zweig
Feuerbach Ewald Eliot Vergil
Goethe Elisabeth von Österreich London
Mendelssohn Balzac Shakespeare Dostojewski Ganghofer
Trackl Lichtenberg Rathenau Doyle Gjellerup
Stevenson Hambruch
Mommsen Tolstoi Lenz Droste-Hülshoff
Thoma von Arnim Hanrieder
Dach Verne Hägele Hauff Humboldt
Reuter Rousseau Hagen Hauptmann Gautier
Karrillon Garschin Defoe Baudelaire
Damaschke Descartes Hebbel
Hegel Kussmaul Herder
Wolfram von Eschenbach Dickens Schopenhauer Rilke George
Darwin Melville Grimm Jerome
Bronner Bebel Proust
Campe Horváth Aristoteles Voltaire Federer Herodot
Bismarck Vigny Barlach
Gengenbach Heine
Storm Casanova Tersteegen Grillparzer Georgy
Chamberlain Lessing Langbein Gilm Gryphius
Brentano Lafontaine
Strachwitz Claudius Schiller Kralik Iffland Sokrates
Katharina II. von Rußland Bellamy Schilling
Gerstäcker Raabe Gibbon Tschechow
Löns Hesse Hoffmann Gogol Wilde Vulpius
Luther Heym Hofmannsthal Gleim
Roth Klee Hölty Morgenstern Goedicke
Heyse Klopstock Kleist
Luxemburg Puschkin Homer Mörike
La Roche Horaz Musil
Machiavelli Kierkegaard Kraft Kraus
Navarra Aurel Musset
Nestroy Marie de France Lamprecht Kind Kirchhoff Hugo Moltke
Laotse Ipsen Liebknecht
Nietzsche Nansen Ringelnatz
Marx Lassalle Gorki Klett Leibniz
von Ossietzky May vom Stein Lawrence Irving
Petalozzi Knigge
Platon Pückler Michelangelo Kafka
Sachs Poe Liebermann Kock
de Sade Praetorius Mistral Zetkin Korolenko

The publishing house tredition has created the series **TREDITION CLASSICS**. It contains classical literature works from over two thousand years. Most of these titles have been out of print and off the bookstore shelves for decades.

The book series is intended to preserve the cultural legacy and to promote the timeless works of classical literature. As a reader of a **TREDITION CLASSICS** book, the reader supports the mission to save many of the amazing works of world literature from oblivion.

The symbol of **TREDITION CLASSICS** is Johannes Gutenberg (1400 – 1468), the inventor of movable type printing.

With the series, tredition intends to make thousands of international literature classics available in printed format again – worldwide.

All books are available at book retailers worldwide in paperback and in hardcover. For more information please visit: www.tredition.com

tredition was established in 2006 by Sandra Latusseck and Soenke Schulz. Based in Hamburg, Germany, tredition offers publishing solutions to authors and publishing houses, combined with worldwide distribution of printed and digital book content. tredition is uniquely positioned to enable authors and publishing houses to create books on their own terms and without conventional manufacturing risks.

For more information please visit: www.tredition.com

Wild Beasts and Their Ways, Reminiscences of Europe, Asia, Africa and America —Volume 1

Samuel White, Sir Baker

Imprint

This book is part of the TREDITION CLASSICS series.

Author: Samuel White, Sir Baker
Cover design: toepferschumann, Berlin (Germany)

Publisher: tredition GmbH, Hamburg (Germany)
ISBN: 978-3-8491-5464-6

www.tredition.com
www.tredition.de

CHAPTER I

THE RIFLE OF A PAST HALF CENTURY

Forty years ago our troops were armed with a smooth-bore musket, and a small force known as the "Rifle Brigade" was the exception to this rule.

The military rifle carried a spherical bullet, and, like all others of the period, it necessitated the use of a mallet to strike the ball, which, being a size larger than the bore, required the blow to force it into the rifling of the barrel in order to catch the grooves.

Sporting rifles were of various sizes, but they were constructed upon a principle generally accepted, that extreme accuracy could only be obtained by burning a very small charge of powder.

The outfit required a small mallet made of hardwood faced with thick buff leather, a powerful loading-rod, a powder-flask, a pouch to contain greased linen or silk patches; another pouch for percussion caps; a third pouch for bullets. In addition to this cumbersome arrangement, a nipple-screw was carried, lest any stoppage might render necessary the extraction of the nipple.

The charge of powder in ordinary use for a No. 16 bore (which carried an ounce spherical ball) was 1 1/2 dram, and the sights were adjusted for a maximum range of 200 yards. Although at this distance considerable accuracy could be attained at the target upon a quiet day, it was difficult to shoot with any precision at an unmeasured range owing to the high trajectory of the bullet. Thus for sporting purposes it was absolutely essential that the hunter should be a first-rate judge of distance in order to adjust the sights as required by the occasion. It was accordingly rare to meet with a good rifle-shot fifty years ago. Rifle-shooting was not the amusement sought by Englishmen, although in Switzerland and Germany it was the ordinary pastime. In those countries the match-rifle was immensely heavy, weighing, in many instances, 16 lbs., although the bullet was exceedingly small.

The idea of non-recoil was paramount as necessary to ensure accuracy.

It will be at once perceived that the rifle was a most inferior weapon, failing through a low velocity, high trajectory, and weakness of penetration.

In 1840, I had already devoted much attention to this subject, and I drew a plan for an experimental rifle to burn a charge of powder so large that it appeared preposterous to the professional opinions of the trade. I was convinced that accuracy could be combined with power, and that no power could be obtained without a corresponding expenditure of powder. Trajectory and force would depend upon velocity; the latter must depend upon the volume of gas generated by explosion.

The rifle was made by Gibbs of Bristol. The weight was 21 lbs., length of barrel 36 inches, weight of spherical belted bullet 3 ounces, of conical bullet 4 ounces, charge of powder 16 drams. The twist was one full turn in the length of barrel. The rifling was an exceedingly deep and broad groove (two grooves), which reduced the difficulty of loading to a minimum, as the projecting belt enabled the bullet to catch the channel instantly, and to descend easily when wrapped in a greased silk patch without the necessity of hammering. The charge of powder was inserted by inverting the rifle and passing up the loading-rod with an ounce measure screwed to the end; this method prevented the powder from adhering to the sides of the barrel, and thus fouling the grooves.

An extraordinary success attended this rifle, which became my colossal companion for many years in wild sports with dangerous game. It will be observed that the powder charge was one-third the weight of the projectile, and not only a tremendous crushing power, but an extraordinary penetration was obtained, never equalled by any rifle that I have since possessed.

This weapon was in advance of the age, as it foreshadowed the modern Express, and the principle was thoroughly established to my own satisfaction, that a sporting rifle to be effective at a long range must burn a heavy charge of powder, but the weight of the weapon should be in due proportion to the strain of the explosion.

When I first visited Ceylon in 1845, there were several renowned sportsmen who counted their slain elephants by many hundreds, but there were no rifles. Ordinary smooth-bore shot-guns were the favourite weapons, loaded invariably with a double charge of powder and a hardened ball. In those days the usual calibre of a gun was No. 14 or 16. A No. 12 was extremely rare. The charge for No. 16 was 2 3/4 drams of fine grain powder, and drams for No. 12. Accordingly, the light guns, or "fowling-pieces," as they were termed, were severely tested by a charge of 6 drams of the strongest powder with a hardened bullet; nevertheless I never heard of any failure.

At a short range the velocity and penetration of an ounce spherical ball, with the heavy powder charge, were immense, but beyond 50 yards the accuracy was imperfect.

I believe I was the first to introduce rifles into Ceylon, which were then regarded by the highest authorities in the island as impractical innovations, too difficult to sight, whereas an ordinary gun could be used with ball more quickly in taking a snap-shot.

The rifles which I had provided were heavy, the 3 ounce already mentioned, 21 lbs., and a long 2 ounce by Blisset, 16 lbs. The latter was a polygroove, the powder charge only 1 1/2 dram when I originally purchased it. It was wonderfully accurate at short ranges with the small charge, which I quickly increased to 6 drams, thereby losing accuracy, but multiplying velocity.

Twelve months' experience with elephants and buffaloes decided me to order a battery of double-barrelled rifles, No. 10, two-grooved, with 6 drams of fine grain powder, and spherical-belted bullets. These were most satisfactory, and they became the starting-point for future experiments.

Shortly before the Crimean War, the musket was abolished, and about 1853 the British army was armed throughout with rifles. The difficulty of a military rifle lay in the rapid fouling of the barrel, which necessitated a bullet too small to expand sufficiently to fill the grooves; this resulted in inaccuracy. Even if the bullet were properly fitted, it became impossible to load when the barrel began to foul after a few discharges.

At that time I submitted a plan to the authorities which simplified the difficulty, and having left the pattern bullet at Woolwich, it quickly appeared with a slight modification as the "Boxer bullet." My plan designed a cone hollowed at the base. The bullet was a size smaller than the bore, which enabled it to slide easily down the barrel when foul. The hollow base fitted upon a cone of boxwood pointed at the insertion, but broad at the base, which was larger than the diameter of the hollow in the bullet. It may be easily understood that although this compound bullet was smaller than the bore of the rifle, a blow with the ramrod after loading would drive the conical bullet upon the larger diameter of the boxwood cone, which, acting like a wedge, would expand the lead, thus immediately secured within the barrel. The expansion when fired drove the boxwood into the centre of the bullet, which of necessity took the rifling.

The Boxer bullet superseded the boxwood plug by the use of a piece of burnt clay, which was less expensive and equally serviceable.

Before breechloaders were invented, we were obliged to fit out a regular battery of four double rifles for such dangerous game as elephants, buffaloes, etc., as the delay in re-loading was most annoying and might lead to fatal accidents.

In hot damp climates it became necessary to fire off and clean the entire battery every evening, lest a miss-fire should be the consequence upon the following morning from the condensation of moisture in the nipple during night. This was not only great trouble and a wasteful expenditure of ammunition, but the noise of so many loud reports just at the hour when wild animals were on the move, alarmed the country. Trustworthy gun-carriers are always difficult to procure, and it was by no means uncommon that in moments of danger, when the spare rifles were required, the gun-bearers had bolted from the scene, and the master was deserted.

The introduction of breechloaders has made shooting a luxury, and has obviated the necessity of a large battery of guns. For military purposes the breechloader has manifold advantages—as the soldier can load while lying down, and keep up a rapid fire from a secure cover. It was remarked during the Crimean War that a large proportion of wounded men were struck in the right arm, which would have been raised above the head when loading the old-fashioned rifle, and was thus prominently exposed.

It is not my intention to enter into the minutiae of military rifles, but I cannot resist the satisfaction with which I regard the triumph of the small-bore which I advocated through the columns of the Times in 1865, at a time when the idea was opposed by nearly all authorities as impracticable, owing to the alleged great drawback of rapid fouling. There can be no doubt that the charge of 70 grains with a small-bore bullet, '303, will have a lower trajectory (higher velocity (equivalent to long range)) than a heavier projectile, '450, with the additional advantage of a minimum recoil.

The earliest in the field of progress was the old-established firm of Purdey and Co. Mr. Purdey, before the general introduction of breechloaders, brought out an Express rifle, No. 70 bore, with a mechanically fitting two-groove solid bullet. This small projectile was a well-pointed cone weighing exactly 200 grains, with a powder charge of 110 grains, more than half the weight of the bullet. The extremely high velocity of this rifle expanded the pure soft lead upon impact with the skin and muscles of a red deer. At the same time there was no loss of substance in the metal, as the bullet, although much disfigured, remained intact, and continued its course of penetration, causing great havoc by its increased surface. Nothing has surpassed this rifle in velocity, although so many improvements have taken place since the introduction of breechloaders, but in the days of muzzle-loaders it was a satisfaction to myself that I was the first to commence the heavy charge of powder with the 3 ounce bullet and 16 drams, to be followed after many years by so high an authority as Mr. Purdey with a 200 grain bullet and 110 grains of powder, thus verifying the principle of my earliest experience.

This principle is now universally accepted, and charges of powder are used, as a rule, which forty years ago would have been regarded as impossible.

The modern breechloader in the hands of a well-trained soldier should be a most deadly weapon, nevertheless we do not find a greater percentage of destruction among the numbers engaged than resulted from the old Brown Bess. The reason is obvious: battles are now fought at long ranges, whereas in the early portion of the century fire was seldom opened at a greater distance than 200 yards, and the actual struggle terminated at close quarters.

A long-range rifle in the excitement of a hot action has several disadvantages. The sights may have been set for 600 or 800 yards when the enemy was at a distance, but should that interval be decreased by an approach at speed, the sights would require an immediate readjustment, otherwise the bullets would fly overhead, and the nearer the enemy advanced, the safer he would be. Troops require most careful training with the new weapons entrusted to their care. Although a rapidity of fire if well directed must have a terrible result, there can be no question that it engenders a wild excitement, and that a vast amount of ammunition is uselessly expended, which, if reserved by slower but steady shooting, would be far more deadly.

Although the difficulty is great in preventing troops from independent firing when their blood is up in the heat of combat, the paramount duty of an officer should be to control all wildness, and to insist upon volleys in sections of companies by word of command, the sights of the rifles being carefully adjusted, and a steady aim being taken at the knees of the enemy.

There cannot be a better example than the advice upon this subject given by the renowned General Wolfe (who was subsequently killed at the siege of Quebec) to the 20th Regiment, of which he was Colonel, when England was hourly expecting an invasion by the French:— … "There is no necessity for firing very fast; … a cool well-levelled fire with the pieces carefully loaded is much more destructive than the quickest fire in confusion."—At Canterbury, 17th December 1755.

This instruction should be sternly impressed upon the minds of all soldiers, as it is the text upon which all admonitory addresses should be founded. It must not be forgotten that General Wolfe's advice was given to men armed with the old muzzle-loading Brown

Bess (musket), which at that time was provided with a lock of flint and steel. Notwithstanding the slowness of fire necessitated by this antiquated weapon, the General cautioned his men by the assurance, "There is no necessity for firing very fast," etc., etc.

The breechloader is valuable through the power which exists, especially with repeating rifles, for pouring in an unremitting fire whenever the opportunity may offer, but under ordinary circumstances the fire should be reserved with the care suggested by the advice of General Wolfe.

Small-bores have become the fashion of the day, and for military purposes they are decidedly the best, as a greater amount of ammunition can be carried by the soldier, while at the same time the range and trajectory of his weapon are improved. The new magazine rifle adopted by the Government is only '303, but this exceedingly small diameter will contain 70 grains of powder with a bullet of hard alloy weighing 216 grains.

For sporting purposes the small-bore has been universally adopted, but I cannot help thinking that like many other fashions, it has been carried beyond the rules of common sense.

When upon entering a gunmaker's shop the inexperienced purchaser is perplexed by the array of rifles and guns, varying in their characters almost as much as human beings, he should never listen to the advice of the manufacturer until he has asked himself what he really requires.

There are many things to be considered before an order should be positively given. What is the rifle wanted for? What is the personal strength of the purchaser? In what portion of the world is he going to shoot? Will he be on foot, or will he shoot from horseback or from an elephant? Will the game be dangerous, or will it be confined to deer, etc.?

Not only the weapon but the ammunition will depend upon a reply to these questions, and the purchaser should strongly resist the delusion that any one particular description will be perfect as a so-called general rifle. You may as well expect one kind of horse or one pattern of ship to combine all the requirements of locomotion as to suppose that a particular rifle will suit every variety of game or condition of locality.

In South Africa accuracy is necessary at extremely long ranges for the open plains, where antelopes in vast herds are difficult of approach. In Indian jungles the game is seldom seen beyond fifty or sixty yards. In America the stalking among the mountains is similar to that of the Scottish Highlands, but upon a larger scale. In Central Africa the distances are as uncertain as the quality of the animals that may be encountered.

Upon the level plains of India, where the blackbuck forms the main object of pursuit, extreme accuracy and long range combined are necessary, with a hollow Express bullet that will not pass through the body. How is it possible that any one peculiar form of rifle can combine all these requirements? Rifles must be specially adapted for the animals against which they are to be directed. I have nothing to do with the purse, but I confine my remarks to the weapons and the game, and I shall avoid technical expressions.

The generally recognised small-bores, all of which are termed "Express" from the large charge of powder, are as follow:—

Small-bore Charge of Large- Charge of For all Game
Express. Powder. bores. Powder. such as*

'577 6 1/2 drams 4 bore 14 drams Elephants. '500 5 1/2 " 8 " 14 " Rhinoceros. '450 5 " 10 " 12 " Buffaloes. '400 4 " 12 " 10 " '360 Toys. '295 Toys.

The two latter rifles, '360 and '295, are charming additions, and although capable of killing deer are only to be recommended as companions for a stroll but not to be classed as sporting rifles for ordinary game. They are marvellously accurate, and afford great satisfaction for shooting small animals and birds. The '360 may be used for shooting black-buck, but I should not recommend it if the hunter possesses a '400.

It would be impossible to offer advice that would suit all persons. I can therefore only give a person opinion according to my own experience.

For all animals above the size of a fallow deer and below that of a buffalo I prefer the '577 solid Express—648 grains solid bullet,—6 drams powder not 6 1/2, as the charge of only 6 drams produces greater accuracy at long ranges.

The weight of this rifle should be 11 1/2 lbs., or not exceeding 12 lbs. For smaller game, from fallow deer downwards, I prefer the '400 Express with a charge of from 85 grains to 4 drams of powder—solid bullet, excepting the case of black-buck, where, on account of numerous villages on the plains, it is necessary that the bullet should not pass through the body. The important question of weight is much in favour of the '400, as great power and velocity are obtained by a weapon of only 8 1/2 lbs.

I should therefore limit my battery to one '577, one '400, and one Paradox No. 12, for ordinary game in India, as elephants and other of the larger animals require special outfit. The Paradox*, invented by Colonel Fosberry and manufactured by Messrs. Holland and Holland of Bond Street, is a most useful weapon, as it combines the shot-gun with a rifle that is wonderfully accurate within a range of 100 yards. (* Since this was written Messrs. Holland have succeeded after lengthened experiments in producing a Paradox No. 8, which burns 10 drams of powder, and carries a very heavy bullet with extreme accuracy. This will be a new departure in weapons for heavy game.)

It is a smooth-bore slightly choked, but severely rifled for only 1 1/2 inch in length from the muzzle. This gives the spin to the projectile sufficient to ensure accuracy at the distance mentioned.

The No. 12 Paradox weighs 84 lbs. and carries a bullet of 1 3/4 ounce with 4 1/2 drams of powder. Although the powder charge is not sufficient to produce a high express velocity, the penetration and shock are most formidable, as the bullet is of hardened metal, and it retains its figure even after striking a tough hide and bones. The advantage of such a gun is obvious, as it enables a charge of buck-shot to be carried in the left barrel, while the right is loaded with a heavy bullet that is an admirable bone-smasher; it also supersedes the necessity of an extra gun for small game, as it shoots No. 6 shot with equal pattern to the best cylinder-bored gun.

There are many persons who prefer a '500 or a '450 Express to the '577 or the '400. I have nothing to say against them, but I prefer those I have named, as the '577 is the most fatal weapon that I have ever used, and with 6 or 6 1/2 drams of powder it is quite equal to any animal in creation, provided the shot is behind the shoulder. This provision explains my reason for insisting that all animals from a buffalo upwards should be placed in a separate category, as it is frequently impossible to obtain a shoulder shot, therefore the rifles for exceedingly heavy game must be specially adapted for the work required, so as to command them in every conceivable position.

I have shot with every size of rifle from a half pounder explosive shell, and I do not think any larger bore is actually necessary than a No. 8, with a charge of 12 or 14 drams of powder. Such a rifle should weigh 15 lbs., and the projectile would weigh 3 ounces of hardened metal.

The rifles that I have enumerated would be always double, but should the elephant-hunter desire anything more formidable, I should recommend a single barrel of 36 inches in length of bore, weighing 22 lbs., and sighted most accurately to 400 yards. Such a weapon could be used by a powerful man from the shoulder at the close range of fifty yards, or it could be fired at long ranges upon a pivot rest, which would enable the elephant-hunter to kill at a great distance by the shoulder shot when the animals were in deep marshes or on the opposite side of a river. I have frequently seen elephants in such positions when it was impossible to approach within reasonable range. A rifle of this description would carry a half-pound shell with an exploding charge of half an ounce of fine grain powder and the propelling charge would be 16 drams. I had a rifle that carried a similar charge, but unfortunately it was too short, and was only sighted for 100 yards. Such a weapon can hardly be classed among sporting rifles, but it would be a useful adjunct to the battery of a professional hunter in Africa.

There can be little doubt that a man should not be overweighted, but that every person should be armed in proportion to his physical strength. If he is too light for a very heavy rifle he must select a smaller bore; if he is afraid of a No. 8 with 14 drams, he must be content with a No. 12 and 10 drams, but although he may be successful with the lighter weapon, he must not expect the performance will equal that of the superior power.

It may therefore be concluded that for a man of ordinary strength, the battery for the heaviest game should be a pair of double No. 8 rifles weighing 14 or 15 lbs. to burn from 12 to 14 drams of powder, with a hardened bullet of 3 ounces. Such a rifle will break the bones of any animal from an elephant downwards, and would rake a buffalo from end to end, which is a matter of great importance when the beast is charging.

Although the rifle is now thoroughly appreciated, and sportsmen of experience have accepted the Express as embodying the correct principle of high velocity, I differ with many persons of great authority in the quality of projectiles, which require as much consideration as the pattern of the gun.

The Express rifle is a term signifying velocity, and this is generally accompanied by a hollow bullet which is intended to serve two purposes— to lighten the bullet, and therefore to reduce the work of the powder, and to secure an expansion and smash-up of the lead upon impact with the animal. I contend that the smashing up of the bullet is a mistake, excepting in certain cases such as I have already mentioned, where the animal is small and harmless like the blackbuck, which inhabits level plains in the vicinity of population, and where the bullet would be exceedingly dangerous should it pass through the antelope and ricochet into some unlucky village.

As I have already advised the purchaser of a rifle to consider the purpose for which he requires the weapon, in like manner I would suggest that he should reflect upon the special purpose for which he requires the bullet. He should ask himself the questions—"What is a bullet?" and "What is the duty of a bullet?"

A bullet is generally supposed to be a projectile capable of retaining its component parts in their integrity. The duty of the bullet is to preserve its direct course; it should possess a power of great penetration, should not be easily deflected, and together with penetrating power it should produce a stunning effect by an overpowering striking energy.

How are we to combine these qualities? If the projectile has great penetrating force it will pass completely through an animal, and the striking energy will be diminished, as the force that should have been expended upon the body is expending itself in propelling the bullet after it has passed through the body. This must be wrong, as it is self-evident that the striking energy or knock-down blow must depend upon the resistance which the body offers to the projectile. If the bullet remains within it, the striking energy; complete and entire, without any waste whatever, remains within the body struck. If, therefore, a bullet '577 of 648 grains propelled by 6 drams of powder has at fifty yards a striking energy of 3500 foot pounds, that force is expended upon the object struck,—provided it is stopped by the opposing body.

We should therefore endeavour to prevent the bullet from passing through an animal, if it is necessary to concentrate the full power of the projectile upon the resisting body.

This is one reason adduced in favour of the hollow Express bullet, which smashes up into minute films of lead when it strikes the hard muscles of an animal, owing to its extreme velocity, and the weakness of its parts through the hollowness of its centre.

I contend, on the contrary, that the bullet has committed suicide by destroying itself, although its fragments may have fatally torn and injured the vital organs of the wounded animal. The bullet has ceased to exist, as it is broken into fifty shreds; therefore it is dead, as it is no longer a compact body,—in fact, it has disappeared, although the actual striking energy of a very inferior bullet may have been expended upon the animal.

If the animal is small and harmless, this should be the desired result. If, on the other hand, the animal should be large and dangerous, there cannot be a greater mistake than the hollow Express projectile.

I have frequently heard persons of great experience dilate with satisfaction upon the good shots made with their little '450 hollow Express exactly behind the shoulder of a tiger or some other animal. I have also heard of their failures, which were to themselves sometimes incomprehensible. A solid Express '577 NEVER fails if the direction is accurate towards a vital part. The position of the animal does not signify; if the hunter has a knowledge of comparative anatomy (which he must have, to be a thoroughly successful shot) he can make positively certain of his game at a short distance, as the solid bullet will crash through muscle, bone, and every opposing obstacle to reach the fatal organ. If the animal be a tiger, lion, bear, or leopard, the bullet should have the power to penetrate, but it should not pass completely through. If it should be a wapiti, or sambur stag, the bullet should also remain within, retained in all cases under the skin upon the side opposite to that of entrance. How is this to be managed by the same rifle burning the same charge of powder with a solid bullet?

The penetration must be arranged by varying the material of the bullet. A certain number of cartridges should be loaded with bullets of extreme hardness, intended specially for large thick-skinned animals; other bullets should be composed of softer metal, which would expand upon the resisting muscles but would not pass completely through the skin upon the opposite side. The cartridges would be coloured for distinction.

If the metal is pure lead, the bullet '577, with an initial velocity of 1650 feet per second, will assuredly assume the form of a button mushroom immediately upon impact, and it will increase in diameter as it meets with resistance upon its course until, when expended beneath the elastic hide upon the opposite side, it will have become fully spread like a mature mushroom, instead of the button shape that it had assumed on entrance. I prefer pure lead for tigers, lions, sambur deer, wapiti, and such large animals which are not thick-skinned, as the bullet alters its form and nevertheless remains intact, the striking energy being concentrated within the body.

The difference in the striking energy of a hollow bullet from that of a solid projectile is enormous, owing to the inequality in weight. The hollow bullet wounds mortally, but it does not always kill neatly. I have seen very many instances where the '500 hollow Express with 5 drams of powder has struck an animal well behind the shoulder, or sometimes through the shoulder, and notwithstanding the fatal wound, the beast has galloped off as though untouched, for at least a hundred yards, before it fell suddenly, and died.

This is clumsy shooting. The solid bullet of pure lead would have killed upon the spot, as the bullet would have retained its substance although it altered its form, and the shock would have been more severe, The hollow bullet exhibits a peculiar result in a post-mortem examination: the lungs may be hopelessly torn and ragged, the liver and the heart may be also damaged, all by the same projectile, because it has been converted into small shot immediately upon impact. Frequently a minute hole will be observed upon the entrance, and within an inch beneath the skin a large aperture will be seen where an explosion appears to have taken place by the breaking-up of the lead, all of which has splashed into fragments scattering in every direction.

Common sense will suggest that although such a bullet will kill, it is not the sort of weapon to stop a dangerous animal when in full charge. Weak men generally prefer the hollow Express because the rifle is lighter and handier than the more formidable weapon, and the recoil is not so severe, owing to the lightness of the bullet.

My opinion may be expressed in a few words. If you wish the bullet to expand, use soft lead, but keep the metal solid. If you wish for great penetration, use hard solid metal, either 1/10 tin or 1/13 quicksilver. Even this will alter its form against the bones of a buffalo, but either of the above will go clean through a wapiti stag, and would kill another beyond it should the rifle be '577 fired with 6 drams of powder.

The same rifle will not drive a soft leaden solid bullet through a male tiger if struck directly through the shoulder; it will be found flattened to a mushroom form beneath the skin upon the other side, having performed its duty effectively, by killing the tiger upon the spot, and retaining intact the metal of which it was composed.

A post-mortem inquiry in the latter case would be most satisfactory. If the bullet shall have struck fair upon the shoulder-joint, it will be observed that although it has retained its substance, the momentum has been conveyed to every fragment of crushed bone, which will have been driven forward through the lungs like a charge of buckshot, in addition to the havoc created by the large diameter of an expanded '577 bullet. Both shoulders will have been completely crushed, and the animal must of course be rendered absolutely helpless. This is a sine qua non in all shooting. Do not wound, but kill outright; and this you will generally do with a '577 solid bullet of pure lead, or with a Paradox bullet 1 3/4 ounces hard metal and 4 1/2 drams of powder. This very large bullet is sufficiently formidable to require no expansion.

Gunmakers will not advise the use of pure lead for bullets, as it is apt to foul the barrel by its extreme softness, which leaves a coating of the metal upon the surface of the rifling. For military purposes this objection would hold good, but so few shots are fired at game during the day, that no disadvantage could accrue, and the rifle would of course be cleaned every evening.

The accidents which unfortunately so often happen to the hunters of dangerous game may generally be traced to the defect in the rifles employed. If a shooter wishes to amuse himself in Scotland among the harmless red deer, let him try any experiments that may please him; but if he is a man like so many who leave the shores of Great Britain for the wild jungles of the East, or of Africa, let him at once abjure hollow bullets if he seeks dangerous game. Upon this subject I press my opinion, as I feel the immense responsibility of advice should any calamity occur. It is only a few months since the lamented Mr. Ingram was killed by an elephant in the Somali country, through using a '450 Express hollow bullet against an animal that should at least have been attacked with a No. 10. I submit the question to any admirer of the hollow Express. "If he is on foot, trusting only to his rifle for protection, would he select a hollow Express, no matter whether '577, '500, or '450; or would he prefer a solid bullet to withstand a dangerous charge?"

India is a vast empire, and various portions, according to the conditions of localities, have peculiar customs for the conduct of wild sports. In dense jungles, where it would be impossible to see the game if on foot, there is no other way of obtaining a shot except by driving. The gunners are in such case placed at suitable intervals upon platforms called mucharns, securely fitted between convenient forks among the branches of a tree, about 10 or 12 feet above the ground. From this point of vantage the gunner can see without being seen and, thoroughly protected from all danger, he may amuse himself by comparing the success of his shooting with the hollow Express or with the solid bullet at the animals that pass within his range, which means a limit of about 50 yards. I contend that at the short distance named; a tiger should NEVER escape from a solid bullet; he often escapes from the hollow bullet for several reasons.

It must be remembered that animals are rarely seen distinctly in a thick jungle, countless twigs and foliage intercept the bullet, and the view, although patent to both open eyes, becomes misty and obscure when you shut one eye and squint along the barrel. You then discover that although you can see the dim shadow of your game, your bullet will have to cut its way through at least twenty twigs before it can reach its goal. A solid bullet may deflect slightly, but it will generally deliver its message direct, unless the opposing objects are more formidable than ordinary small branches. A hollow bullet from an Express rifle will fly into fragments should it strike a twig the size of the little finger. This is quite sufficient to condemn the hollow projectile without any further argument.

While writing the above, I have received the Pioneer, 24th June 1888, which gives the following account of an escape from a tiger a few weeks ago by Mr. Cuthbert Fraser, and no better example could be offered to prove the danger of a hollow bullet. It will be seen that a solid bullet would have killed the tiger on the spot, as it would have penetrated to the brain, instead of which it broke into the usual fragments when striking the hard substance of the teeth, and merely destroyed one eye. The bullet evidently splashed up without breaking the jaw, as the wounded animal was not only capable of killing the orderly, but Mr. Fraser "heard, in fact, the crunching of the man's bones." He says "that he felt that he had the tiger dead when he fired, but the Express bullet unfortunately broke up." He had fired the left-hand barrel into the tiger's chest without the slightest result in checking the onset; had that been a solid bullet it would have penetrated to the heart or lungs.

ADVENTURE WITH A TIGER.

The following experience of a sportsman in the Deccan is from the Secunderabad paper of 14th June 1888: —

"Mr. Cuthbert Fraser had a most miraculous escape from a tiger the other day at Amraoti. The lucky hero of this adventure is a District Superintendent of Police in Berar. He is well remembered in Secunderabad as Superintendent of the Cantonment Police before Mr. Crawford. A son of Colonel Hastings Fraser, one of the Frasers of Lovat, he has proved his possession of that nerve and courage which rises to the emergency of danger — on which qualities more than all else the British Empire in India has been built, and on which, after all is said, in the last resort, it must be still held to rest. To quote the graphic account of a correspondent, the escape was about as narrow as was ever had. Mr. Fraser was told by his orderly that the tiger was lying dead with his head on the root of a tree. The orderly having called him up, he went to the spot. Mr. Fraser then sent the orderly and another man with the second gun back, and knelt down to look. Just then the tiger roared and came at him from about eighteen feet off: he waited till the tiger was within five feet of him and fired. As the tiger did not drop, he fired his second shot hurriedly. The first shot had hit exactly in the centre of the face but just an inch too low. It knocked the tiger's right eye out and smashed all the teeth of that side of the jaw. The second shot struck the tiger in the chest, but too low. What happened then Mr. Fraser does not exactly know, but he next found himself lying in front of the tiger, one claw of the beast's right foot being hooked into his left leg, in this way trying to draw Mr. Fraser towards him; the other paw was on his right leg. Mr. Fraser's chin and coat were covered with foam from the beast's mouth. He tried hard to draw himself out of the tiger's clutches. Fortunately the beast was not able to see him, as Mr. Fraser was a little to one side on the animal's blind side and the tiger's head was up. Suddenly seeing Mr. Fraser's orderly bolting, he jumped up and went for the man, and catching him he killed him on the spot. Mr. Fraser had lost his hat, rifle, and all his cartridges, which had tumbled out of his pocket. He jumped up, however, and ran to the man who had his second gun, and to do so had to go within eight paces of the spot where the tiger was crouching over his orderly. He heard, in fact, the crunching of the man's bones and saw the tiger biting the back of the head. He now took the gun from his man. The latter said that he had fired both barrels into the tiger — one when he was crouching over Mr. Fraser, and the other when he was over the prostrate body of the orderly. The man had fired well and true, but just too far back, in his anxiety not to hit the man he would save, instead of the tiger. When afterwards asked if he was not afraid to hit the Sahib, 'I was very much afraid indeed,' he replied, 'but dil mazbut karke lagaya: I nerved myself for the occasion.' 'A good man and true!' a high officer writes, 'who after firing never moved an inch till Mr. Fraser came to him, although close to the tiger all the while. He is one of the Gawilghur Rajputs — a brave race, Ranjit Singh, a good name.' The man said he had no more cartridges left and so they both got a little farther from the tiger, as the orderly was evidently done for. Afterwards they found one more cartridge for the gun and tried to recover the body, but it was no use. The tiger was lying close, most of the buffaloes had bolted and the Kurkoos would not help. Mr. Fraser then sent six miles off for an elephant. But the animal did not arrive till dark, so Mr. Fraser went home in great grief about the poor orderly and at having to leave the body. His own wound was bleeding a great deal, it being a deep claw gash. Next day they got the body and the tiger dead, lying close to each other. Perhaps no narrower escape than Mr. Fraser's has ever been heard of. To the excellent shot which knocked the beast's eye out he undoubtedly owes his life. He says that he felt that he had the tiger dead when he fired, but the Express bullet unfortunately broke up. Probably, he thinks a 12-bore would have reached the brain."

I could produce numerous instances where failures have occurred, and I know sportsmen of long experience who have given up the use of hollow bullets except against such small game as black-buck and other antelopes or deer.

So much for the Express hollow bullet, after which it is at the option of all persons to please themselves; but personally I should decline the company of any friend who wished to join me in the pursuit of dangerous game if armed with such an inferior weapon. In another portion of this volume I shall produce a striking instance of the result.

The magazine rifle, which is destined to become the military arm of the future, can hardly merit a place among sporting rifles, as it must always possess the disadvantage of altering its balance as the ammunition is expended. The Winchester Company have, I believe, produced a great improvement in a rifle of this kind, '400, which carries a charge of 110 grains of powder; but even so small a bore must be unhandy if the rifle is arranged to contain a supply of cartridges. For my own use I am quite contented with one '577, a '400, and a No. 12 Paradox - all solid bullets, but varying in hardness of metal according to the quality of game; for the largest animals a pair of No. 8 rifles with hard bullets and 14 drams of powder.

I can say nothing more concerning rifles for the practical use of sportsmen, although a volume might be devoted to their history and development. Shot guns are too well understood to merit a special notice.

CHAPTER II

THE ELEPHANT (ELEPHAS)

This animal has interested mankind more than any other, owing to the peculiar combination of immense proportions with extraordinary sagacity. The question has frequently been raised "Whether the elephant or the dog should be accepted as superior in intelligence?" My own experience would decide without hesitation—The Dog is man's companion; the Elephant is his slave.

We all know the attachment and fidelity of the dog, who appears to have been created specially to become the friend of the human race. He attaches himself equally to the poor man and the rich, and shares our fortunes "for better, for worse," clinging with heroic loyalty to his master when all other friends may have abandoned him. The power of memory is wonderfully exhibited, considering the shortness of life which Nature, by some mischance has accorded to man's best friend.

"While thus Florinda spake, the dog who lay Before Rusilla's feet, eyeing him long And wistfully, had recognised at length, Changed as he was and in those sordid weeds, His royal master. And he rose and lick'd His withered hand, and earnestly looked up With eyes whose human meaning did not need The aid of speech; and moan'd, as if at once To court and chide the long-withheld caress... . . . . Disputing, he withdrew. The watchful dog Followed his footsteps close. But he retired Into the thickest grove; there yielding way To his o'erburthen'd nature, from all eyes Apart, he cast himself upon the ground, And threw his arms around the dog, and cried While tears stream'd down. Thou Theron, thou hast known Thy poor lost master... Theron, only thou!"—

Southey's "Roderick, last of the Goths."

In case of danger the dog will defend his master, guided by his own unaided intelligence; he at once detects and attacks the enemy. In wild sports he *shares the delight of hunting equally with his master, and the two are inseparable allies. The day is over, and he lies down and sleeps before the fire at his master's feet, and dreams of the dangers and exploits; he is a member of his master's household.

The elephant is, in my opinion, overrated. He can be educated to perform certain acts, but he would never volunteer his services. There is no elephant that I ever saw who would spontaneously interfere to save his master from drowning or from attack. An enemy might assassinate you at the feet of your favourite elephant, but he would never attempt to interfere in your defence; he would probably run away, or remain impassive, unless guided and instructed by his mahout. This is incontestable; the elephant will do nothing useful unless he is specially ordered to perform a certain work or movement.

While condemning this apathetic character, we must admit that in the elephant the power of learning is extraordinary, and that it can be educated to perform wonders; but such performances are only wonderful as proving the necessary force of direction and guidance by a superior power, to which the animal is amenable.

I have had very many years' experience with elephants, both Asiatic and African, and in my opinion they are naturally timid. Although in a wild state the males are more or less dangerous, especially in Africa, the herd of elephants will generally retreat should they even wind an unseen enemy. This timidity is increased by domestication, and it is difficult to obtain an elephant sufficiently staunch to withstand the attack of any wild animal. They will generally turn tail, and not only retreat gracefully, but will run in a disgraceful panic, to the great danger of their riders should the locality be forest.

The difference in species is distinct between the Asiatic and the African. It is at all times difficult to give the measurement of a dead animal, especially when so enormous, as the pressure of weight when alive would reduce the height afforded by measurement when the body is horizontal.

The well-known African elephant Jumbo that was sold to America by the Zoological Society of London, was brought up in confinement since its early existence, when it was about 4 feet 6 inches high. That elephant was carefully weighed and measured before it left England, with the result, of height at shoulder, 11 feet; weight, six tons and a half. The girth of the fore-foot when the pressure of the animal's weight was exerted, was exactly half the perpendicular height of the elephant. I have seen very much larger animals in Africa, but there is nothing in India to approach the size of Jumbo.

There is no reason why the African elephants should not be tamed and made useful, but the difficulty lies in obtaining them in any great numbers. The natives of Africa are peculiarly savage, and their instincts of destruction prevent them from capturing and domesticating any wild animals. During nine years' experience of Central Africa I never saw a tamed creature of any kind, not even a bird, or a young antelope in possession of a child. The tame elephant would be especially valuable to an explorer, as it could march through streams too deep for the passage of oxen, and in swimming rivers it would be proof against the attacks of crocodiles. So few African elephants have been tamed in proportion to those of Asia that it would be difficult to pronounce an opinion upon their character when domesticated, but it is generally believed by their trainers that the Indian species is more gentle and amenable to discipline. The power of the African is far in excess of the Asiatic. Nine feet at the highest portion of the back is a good height for an Indian male, and eight feet for the female, although occasionally they are considerably larger. There are hardly any elephants that measure ten feet in a direct perpendicular, although the mahouts pretend to fictitious heights by measuring with a tape or cord from the spine, including the curve of the body.

As Jumbo was proved to have attained the height of eleven feet although in captivity from infancy, it may be easily imagined that in a wild state the African elephant will attain twelve feet, or even more. I have myself seen many animals that would have exceeded this, although it would be impossible to estimate their height with accuracy.

The shape of the African variety is very peculiar, and differs in a remarkable manner from the Asiatic. The highest point is the shoulder, and the back is hollow; in the Indian the back is convex, and the shoulder is considerably lower. The head of the African is quite unlike that of the Indian; and the ears, which in the former are enormous, completely cover the shoulder when thrown back. The best direction for a vital shot at an African elephant is at the extremity of the ear when flapped against the side. A bullet thus placed will pass through the centre of the lungs. The Indian elephant has many more laminae in the teeth than the African, constituting a larger grinding surface, as the food is different. The African feeds upon foliage and the succulent roots of the mimosa and other trees, which it digs up with its powerful tusks; the forests are generally evergreen, and being full of sap, the bark is easier to masticate than the skeleton trees of India during the hottest season. Both the Indian and African varieties have only four teeth, composed of laminae of intensely hard enamel, divided by a softer substance which prevents the surface from becoming smooth with age; the two unequal materials retain their inequality in wear, therefore the rough grinding surface is maintained notwithstanding the work of many years. A gland at the posterior of the jaw supplies a tooth-forming matter, and the growth of fresh laminae is continuous throughout life; the younger laminae form into line, and march forward until incorporated and solidified in the tooth.

It is impossible to define exactly the limit of old age, as there can be little doubt that captivity shortens the duration of life to a great degree. We can only form an opinion from the basis of growth when young. As an elephant cannot be fully developed in the perfection of ivory until the age of forty, I should accept that age in a wild animal as the period of a starting-point in life, and I should imagine that the term of existence would be about a hundred and fifty years.

The life of an elephant in captivity is exactly opposed to its natural habits. A wild Indian elephant dreads the sun, and is seldom to be found exposed in the open after dawn of day. It roams over the country in all directions during night, and seeks the shelter of a forest about an hour before the sun rises. It feeds heartily, but wastefully, tearing down branches, half of which it leaves untouched; it strips the bark off those trees which it selects as tasteful, but throws wilfully away a considerable portion. Throughout the entire night the elephant is feeding, and it is curious to observe how particular this animal is in the choice of food. Most wild animals possess a certain amount of botanical knowledge which guides them in their grazing; the only exception is the camel, who would poison himself through sheer ignorance and depraved appetite, but the elephant is most careful in its selection of all that is suitable to its requirements. It is astonishing how few of the forest trees are attractive to this animal. Some are tempting from their foliage, others from their bark (vide the powerfully astringent Catechu), some from the succulent roots, and several varieties from the wood, which is eaten like the sugar-cane. There is one kind of tree the wood of which alone is eaten after the rind has been carefully stripped off.

The elephant, being in its wild state a nocturnal animal, must be able to distinguish the various qualities of trees by the senses of smell and touch, as in the darkness of a forest during night it would be impossible to distinguish the leaves. There are few creatures who possess so delicate a sense of smell; wild elephants will wind an enemy at a distance of a thousand yards, or even more, should the breeze be favourable. The nerves of the trunk are peculiarly sensitive, and although the skin is thick, the smallest substance can be discovered, and picked up by the tiny proboscis at the extremity.

A wound upon any portion of the trunk must occasion intense pain, and the animal instinctively coils the lower portion beneath its chest when attacked by a tiger. This delicacy of nerve renders the elephant exceedingly timid after being wounded, and it is a common and regrettable occurrence that an elephant which has been an excellent shikar animal before it has been injured, becomes useless to face a tiger after it has been badly clawed. I cannot understand the carelessness of an owner who thus permits a good elephant to work unprotected. In ancient days the elephants were armoured for warlike purposes to protect them from spears and javelins, and nothing can be easier than to arrange an elastic protective hood, which would effectually safeguard the trunk and head from the attack of any animal.

I had an excellent hood arranged for a large tusker which was lent to me by the Commissariat. The first layer of material was the soft but thick buff leather of sambur deer. This entirely covered the head, and was laced beneath the throat; at the same time it was secured by a broad leather strap and buckle around the neck. A covering for about three feet from the base of the trunk descended from the face and was also secured by lacing. The lower portion of the trunk was left unprotected, as the animal would immediately guard against danger by curling it up when attacked. Upon this groundwork of buff leather I had plates of thick and hard buffalo hide, tanned, overlapping like slates upon a roof. This armour was proof against either teeth or claws, as neither could hold upon the slippery and yielding hard surface of the leather tiles; at the same time the elephant could move its trunk with ease. Two circular apertures were cut out for the eyes, about six inches in diameter.

An elephant, if well trained, would be sufficiently sagacious to appreciate this protection should it find itself unharmed after a home charge by a tiger or other dangerous beast; and such a quality of armour would add immensely to its confidence and steadiness.

Although the elephant is of enormous strength it is more or less a delicate animal, and is subject to a variety of ailments. A common disease is a swelling in the throat, which in bad cases prevents it from feeding. Another complaint resembles gout in the legs, which swell to a distressing size, and give exquisite pain, especially when touched. This attack is frequently occasioned by allowing elephants, after a long march under a hot sun, to wade belly-deep in cool water in order to graze upon the aquatic vegetation.

Few animals suffer more from the sun's rays than the elephant, whose nature prompts it to seek the deepest shade. Its dark colour and immense surface attract an amount of heat which becomes almost insupportable to the unfortunate creature when forced to carry a heavy load during the hot season in India. Even without a greater weight than its rider, the elephant exhibits signs of distress when marching after 9 a.m. At such times it is disagreeable, as the animal has a peculiar habit of sucking water through the trunk from a supply contained within the stomach, and this it syringes with great force between its fore legs, and against its flanks to cool its sides with the ejected spray. The rider receives a portion of the fluid in his face, and as the action is repeated every five minutes, or less, the operation is annoying.

It is a curious peculiarity in the elephant that it is enabled to suck up water at discretion simply by doubling the trunk far down the throat, and the fluid thus procured has no disagreeable smell, although taken direct from the creature's stomach. In every way the elephant is superior to most animals in the freedom from any unpleasant odour. Its skin is sweet, and the hand retains no smell whatever, although you may have caressed the trunk or any other portion of the body. It is well known that a horse is exceedingly strong in odour, and that nothing is more objectionable than the close proximity of a stable, or even of a large number of horses picqueted in the open,—I have frequently been camped where fifty or sixty elephants were for several days in the same position within a hundred yards of the tents, and still there was no offensive scent.

The food of an elephant is always fresh and clean, and the digestive functions are extremely rapid. The mastication is a rough system of grinding, and the single stomach and exceedingly short intestines simplify the process of assimilation. The rapidity of the food passage necessitates a consumption of a large amount, and no less than six hundred pounds of fodder is the proper daily allowance for an elephant.

There have been frequent discussions upon the important subject of elephant-feeding. Mr. G. P. Sanderson, the superintendent of the keddah department in Assam, has declared against the necessity of allowing a ration of grain in addition to the usual fodder. This must naturally depend upon the quality of the green food. If the locality abounds in plantains, the stems of those plants are eagerly devoured, and every portion except the outside rind is nourishing. Even then the waste is excessive should the stems be heedlessly thrown down before the animal. It will immediately proceed to strip long fibrous ribbons from the stem by placing one foot upon the extremity, and then tearing off the alternate layers like the skin of an onion. These it converts into playthings, throwing them over its back and neck until it is dressed in dangling necklaces, which by degrees, after serving as toys, are ultimately devoured. The proper method of feeding an elephant with plantains where an allowance of rice is added, is by splitting the entire stem through the centre, and then cutting it into transverse sections about two feet in length. As each layer is detached, it resembles a delicately coloured trough, nearly white; this is doubled up in the centre and it at once forms a hollow tube, similar to a very thick drain tile. A handful of rice is placed within, and it is secured by tying with a fibrous strip from the plantain stem. A large pile of these neat packages is prepared for every elephant, and, when ready, the mahout sits by the heap and hands the parcels one by one to the ever-expectant trunk.

The delicacy of an elephant's palate is extraordinary, and the whims of the creature are absurd in the selection or rejection of morsels which it prefers or dislikes. I once saw a peculiar instance of this in an elephant that belonged to the police at Dhubri on the Brahmaputra. This animal had a large allowance of rice, therefore about three-quarters of a pound were placed within each tube of plantain stem. A lady offered the elephant, when being fed, a very small sweet biscuit, about an inch and a half in diameter. This was accepted in the trunk, but almost immediately rejected and thrown upon the ground. The mahout, fearing that his elephant had be-haved rudely in thus refusing a present from a lady's hand, picked up the biscuit and inserted it in the next parcel of rice and plantain stem. This was placed within the elephant's mouth. At the first crunch the animal showed evident signs of disgust, and at once spat out the whole of the contents. There lay a complete ruin of the neat package, which had been burst by the power of the great jaws; but among the scattered rice that had been ejected we perceived the biscuit which had caused the second instance of bad behaviour. So utterly disgusted was the elephant with this tiny foreign substance that it endeavoured to cleanse its mouth from every grain of rice, as though polluted by the contact, and for several minutes it continued to insert its trunk and rake out each atom from its tongue and throat.

The adaptation of the trunk to many purposes is very interesting. I had an elephant who would eat every particle of rice in a round bamboo basket by sucking it up the trunk and then blowing it into its mouth. The basket was close-grained and smooth inside, but although brimful at the commencement of operations, it was emp-tied by the elephant as though it had been cleansed with a dry sponge.

A distinct rule for feeding elephants cannot be laid down without exceptions rendered necessary by peculiarities of localities and the amount of hard work required from the animal. If the elephant is simply turned out to grass for a season, it will thrive upon such natural herbage as bamboos, the foliage of the banyan, peepul, and other varieties of the Ficus family; but if it is expected to travel and perform good work, it is usual in the Commissariat department to allow each elephant seven and a half seers of flour, equal to 15 lbs. avoirdupois. In addition to this, 600 lbs. of green fodder are given, and about 1 lb. of ghee (buffalo butter), with salt and jaggery (native sugar). During a jungle expedition I have always doubled the allowance of flour to 30 lbs. daily for each animal. This is made into large flat cakes like Scotch "scones," weighing 2 lbs. each. The elephants are fed at about an hour before sunset, and then taken to drink water before actual night. Cleanliness is indispensable to the good health and condition of the elephant. It should bathe daily, and the entire body should be well scoured with a piece of brick or a soft quality of sandstone. This operation is much enjoyed, and the huge animal, obeying the command, lies down upon its side and accommodates its carcase to the scrubbing process by adapting its position to the requirements of the operator. It will frequently bury its head completely beneath the water, and merely protrude the extremity of its trunk to breathe above the surface. The coolie is most particular in scrubbing every portion of the animal, after which it will usually stand within the tank or river and shower volumes of water from its trunk over its back and flanks. When well washed, it appears a thoroughly clean black mass, but in a few minutes it proceeds to destroy its personal beauty by throwing clouds of dust upon its back, which, adhering to the moisture occasioned by its recent bath, converts the late clean animal into a brown mound of earth.

There is no quadruped not absolutely amphibious that is so thoroughly at home in the water as the elephant. In a wild state it will swim the largest rivers, and it delights in morasses, where it rolls in the deep mud like a pig or buffalo, and thus coats its hide with a covering of slime, which protects it from the attacks of flies and the worry of mosquitoes. When in a domestic state, the elephant is shy of trusting itself upon unsound earth or quicksands, as it appears to have lost the confidence resulting from an independent freedom among the jungles, and marshy valleys teeming with aquatic vegetation. It will also refuse to cross a bridge unless of solid masonry, and it is curious to observe the extreme care with which it sounds the structure, either by striking with the coiled extremity of the trunk or by experimenting with the pressure of one foot, before it ventures to trust its whole weight upon the suspected floor.

It is difficult to describe the limit of an elephant's swimming powers; this must depend upon many circumstances, whether it is following the stream or otherwise, but the animal can remain afloat for several hours without undue fatigue. The displacement of an elephant's carcase is less than the weight of water, although it swims so deeply immersed that it would appear to float with difficulty. An elephant shot dead within the water will float immediately, with a considerable portion of one flank raised so high above the surface that several men could be supported, as though upon a raft. The body of a hippopotamus will sink like a stone, and will not reappear upon the surface for about two hours, until the gas has to a certain degree distended the carcase: thus the hippopotamus is of a denser and heavier material than the elephant, although it is an aquatic animal.

When tame elephants cross a river they are conducted by their drivers, who stand upon their backs, either balancing themselves without assistance, or supported by holding a cord attached to the animal's neck. It is very interesting to watch the passage of a large river by a herd of these creatures, who to a stranger's eye would appear to be in danger of drowning, although in reality they are merely gamboling in the element which is their delight. I have seen them cross the Brahmaputra when the channel was about a mile in width. Forty elephants scrambled down the precipitous bank of alluvial deposit and river sand: this, although about thirty-five feet high, crumbled at once beneath the fore-foot of the leading elephant, and many tons detached from the surface quickly formed a steep incline. Squatting upon its hind-quarters, and tucking its hinder knees beneath its belly, while it supported its head upon its trunk and outstretched fore legs, it slid and scrambled to the bottom, accompanied by an avalanche of earth and dust, thus forming a good track for the following herd.

It is surprising to see in how few minutes a large herd of elephants descending a steep place will form a road. I have frequently seen them break down an alluvial cliff in the manner described, where at first sight I should have thought it impossible for an elephant to descend. Once within the river the fun began in earnest. After a march in the hot sun, it was delightful to bathe in the deep stream of the Brahmaputra, and the mighty forms splashed and disported themselves, sometimes totally submerged, with the drivers standing ankle-deep upon their hidden backs, which gave them the appearance of walking upon the surface. A tip of the trunk was always above water, and occasionally the animal would protrude the entire head, but only to plunge once more beneath the stream. In this way, swimming at great speed, and at the same time playing along their voyage, the herd crossed the broad river, and we saw their dusky forms glittering in the sunlight as they rose wetted from their bath, and waded majestically along the shallows to reach an island; from which they again started upon a similar journey to cross another channel of the river.

The first impression of a stranger when observing the conduct of a mahout or driver is sympathy for the animal, which is governed through the severe authority of the iron spike. This instrument is about twenty inches long, and resembles somewhat an old-fashioned boat-hook, being a sharp spike at the extremity beyond the keen-pointed hook; it can thus be used either to drive the elephant forward by digging the point into its head, or to pull it back by hooking on to the tender base of the ears. These driving-hooks weigh from about 4 to 6 lbs., and are formidable weapons; some are exceedingly ancient, and have been preserved for a couple of centuries or more, such specimens being highly artistic, and first-rate examples of the blacksmith's work. Although we may commence our experience by pitying the animal that is subjected to such harsh treatment, we quickly discover that without the hook the elephant is like the donkey without the stick. The fact of his knowing that you possess the power, or propeller, is sufficient to ensure comparative obedience, but it would be impossible to direct the movements of an elephant by simple kindness without the power to inflict punishment. This fact alone will prove that the elephant does not serve man through affection, but that it is compelled through fear. It is curious to witness the absurd subjection of this mighty animal even by a child. I have frequently seen a small boy threaten a large elephant with a stick, and the animal has at once winced; and, curling the trunk between the legs, it has closed its eyes and exhibited every symptom of extreme terror when struck repeatedly upon the trunk and face. The male is generally more uncertain than the female. It would at first sight appear that for shooting purposes the bull elephant would be preferred for its greater strength and courage. There can be no doubt that a pair of long tusks is an important protection, and not only forms a defence against the attack of a tiger or other animal, but is valuable for offensive purposes; yet, notwithstanding this advantage, the female is generally preferred to the male, as being more docile and obedient.

The males differ in character, but they are mostly uncertain in temper during a period varying from two to four months every year. At such occurrences of disturbance the animal requires careful treatment, and the chains which shackle the fore legs should be of undoubted quality. Some elephants remain passive throughout the year, while others appear to be thoroughly demented, and, although at other seasons harmless, would, when "must," destroy their own attendant and wreak the direst mischief. At such a crisis the mahout must always be held responsible for accidents, as the animal, if properly watched and restrained, would be incapable of active movements, and would of course be comparatively harmless. Upon many occasions, through the neglect of the attendant, an elephant has been left unchained, or perhaps secured with an old chain that has been nearly worn through a link; the escape of the animal under such circumstances has led to frightful casualties, usually commencing with the destruction of the mahout, who may have attempted a recapture. The approach of the "must" period is immediately perceived by a peculiar exudation of an oily nature from a small duct upon either temple; this somewhat resembles coal-tar in consistence, and it occupies an area of about four inches square upon the surface of the skin. There is a decided odour in this secretion somewhat similar to the same exudation from the neck of the male camel.

I have known male elephants which were remarkably docile throughout all seasons, but even these had to be specially regarded during the period of "must," as there was no means of foretelling a sudden and unexpected outbreak of temper. Many males are at all times fretful, and these expend their ill-nature in various ways; if chained, they kick up the earth, and scatter the dust in all directions; they are never quiet for one moment throughout the day, but continue to swing their heads to and fro, and prick forward their ears, exhibiting a restlessness of spirit that is a sufficient warning to any stranger. Such elephants should always be approached with caution, and never directly in front, but at the side.

An elephant is frequently treacherous, and if the person should stand unheedingly before it, a sudden slap with the trunk might be the consequence. For the same reason, it would be dangerous to approach the heels of such an animal, as a kick from an elephant is rather an extensive movement, and it is extraordinary that so colossal a limb as the hind leg can be projected with such velocity, equalling that of a small pony.

Discussions have frequently arisen concerning the maximum speed of an elephant; this is difficult to decide exactly, as there can be no question that the animal in a wild state will exert a greater speed than can be obtained from it when domesticated. The African variety is decidedly faster than the Asiatic; the legs being longer, the stride is in proportion; and as the habits of the African lead it to wander over large tracts of open country instead of confining its rambles to secluded forests, this peculiarity would naturally render the animal more active, and tend to accelerate its movements. I consider that the African elephant is capable of a speed of fifteen miles an hour, which it could keep up for two or three hundred yards, after which it would travel at about ten miles an hour, and actually accomplish the distance within that period. The Asiatic elephant might likewise attain a speed of fifteen miles for perhaps a couple

of hundred yards, but it would not travel far at a greater pace than eight miles an hour, and it would reduce that pace to six after the first five miles.

The proof of an elephant's power of great speed for a short distance is seldom seen except in cases where the animal is infuriated, and gives chase to some unfortunate victim, who seldom escapes his fate by flight. For a short burst of fifty or one hundred yards an elephant might occasionally attain a pace exceeding fifteen miles an hour, as I have frequently, when among rough ground, experienced a difficulty in escaping when on horseback; and in my young days, when a good runner, I have been almost caught when racing along a level plain as smooth as a lawn with a savage elephant in full pursuit. An active man upon good ground can run for a short distance at the rate of eighteen miles an hour; this should clear him from the attack of most elephants; but unfortunately the good ground is scarce, and the elephant is generally discovered in a position peculiarly favourable to itself, where the roughness of the surface and the tangled herbage render it impossible for a man to run at full speed without falling.

We have recently seen a distressing example in the death of the lamented Mr. Ingram in Somaliland, who, although well mounted, was overtaken by an infuriated wild elephant and killed. This was a female, and it appears that Mr. Ingram, having followed her on horseback, had fired repeatedly with a rifle only .450. The animal charged, and owing to the impediments of the ground, which was covered with prickly aloes, the horse could not escape, and Mr. Ingram was swept off the saddle and impaled upon the elephant's tusks.

The African differs from the Asiatic in the formation of ivory, the tusks of the former being both thicker and heavier; the females also possess tusks, whereas those of the Asiatic variety have merely embryo tusks, which do not project more than two or three inches beyond the lips. I had a tusk of an African elephant that weighed 149 lbs. I have seen in Khartoum a pair that weighed 300 lbs., and I saw a single tusk of 172 lbs. In 1874 a tusk was sold at the ivory sale in London that weighed 188 lbs. These specimens are exceptions to the general rule, as the average weight in a full-grown African male would be about 140 lbs. the pair, or 75 lbs. for one tusk and 65 lbs. for the fellow, which is specially employed for digging.

The African variety is an industrious digger, as it feeds upon the succulent roots of many trees, especially those of the mimosa family. The right tusk is generally used in these operations more than the left; accordingly it is lighter from continual wear, and it is known by the Arabs as the "hadam" or servant. As the African elephant is a root-eater it is far more destructive than the Asiatic. It is astonishing to observe the waste of trees that are upturned by a large herd of these animals, sometimes out of sheer wantonness, during their passage through a forest. The dense tops of mimosas are a great attraction, and there can be no doubt that elephants work collectively to dig out and to overthrow the trees that would be too large for the strength of a single animal. I have seen trees between two and three feet in diameter that have been felled for the sake of the roots and tender heads; these have shown unmistakable signs of an attack by several elephants, as the ground has been ploughed by tusks of different sizes to tear up the long straggling roots which were near the surface, and the deep marks of feet around the centre of operations, of various diameters, have proved the co-operation of members of the herd.

I once saw an elephant strike a large timber tree with its forehead to shake down the fruit. This was a peculiar example of the immense power that can be exerted when required. We were waiting near the margin of the White Nile, about half an hour before sunset, expecting the arrival of waterbuck, when a rumbling sound and a suppressed roar in the jungle were accompanied by the breaking of a branch, which denoted the approach of elephants. Presently they emerged from the forest in several directions, and one, which appeared to be the largest I had ever seen, advanced to within 120 yards of our position without perceiving us, as we were concealed behind a bush upon some rising ground close to the river's bank. This elephant had enormous tusks, but as we had only small-bore rifles, I was contented to watch, without disturbing the magnificent animal before me.

There was a very large and lofty tree quite three feet in diameter; upon the upper branches grew the much-loved fruit, similar in appearance to good-sized dates, and equally sweet and aromatic (Balanites Egyptiaca). Elephants will travel great distances to arrive at a forest where such fruit is produced in quantity, and they appear to know the season when the crop will be thoroughly ripe. Upon this occasion, the elephant, having picked up the single fruits which lay scattered upon the ground, presently looked up, and being satisfied with the appearance of the higher boughs, he determined to shake down a plentiful supply. Retiring for a few feet, he deliberately rammed his forehead against the stem, with such force as to shake the tree from top to bottom, causing a most successful shower of the coveted fruit, which he immediately commenced to eat.

Commander R. N. J. Baker was my companion, and we agreed that any person who might have taken refuge in the branches of that large tree must have held on exceedingly tight to have avoided a fall, so severe was the concussion.

When it is considered that a large bull elephant weighs between six and seven tons, which weight is set in movement by the muscular exertion of the animal, there is at once an explanation of the force against a tree, which, although large, would hardly exceed that weight.

The memory of elephants must be peculiarly keen, as they remember the seasons for visiting certain districts where some particular food is produced in attractive quantities. In the southern district of Ceylon, between Yalle river and the sea-coast, there are great numbers of the Bael tree, the fruit of which resembles a large cricket-ball. The shell is hard, and when ripe it becomes brown, and can only be broken by a sharp blow with some hard substance. The contents are highly aromatic, consisting of a brownish substance exceedingly sweet, and mixed with small seeds resembling those inside a pear. There is a strong flavour of medlar in this fruit, and it is much esteemed for medicinal properties, especially in cases of diarrhoea. Although elephants refuse the Bael fruit unless quite ripe, they will invariably arrive in great numbers during the favourable season in the southern districts of Ceylon. The question arises, "How can an animal remember the month without an almanack?"

There is no doubt that animals possess in many instances a far greater degree of reason than is generally admitted, with which the exercise of memory is so closely allied that it is difficult to separate or define the attributes. An elephant will remember those who have shown kindness, perhaps for a longer period than it will others who may have offended. After seven months' absence in England, an elephant that I had from the Commissariat on my previous visit to India recognised me at once upon my return. I had been in the habit of feeding this animal with sugar-canes and other choice food almost daily during several months' companionship in the jungle; this was not forgotten, and "cupboard love" was harboured in its memory with the expectation that the feeding would be repeated.

In the same manner, but perhaps in a lesser degree, the elephant will remember those whom it dislikes, and during the season of "must" it would be exceedingly dangerous for such persons to venture within reach of the animal's trunk. Stories are numerous concerning the animosity of elephants against their mahouts or other attendants who have cruelly treated them; but, on the other hand, the animals frequently exhibit a wild ferocity towards those who have been innocent of harshness. As characters vary among human beings, and some persons when intoxicated become suddenly brutal, although when sober they have been mild in reputation, so also we find conflicting natures among elephants, and the insane excitement of the "must" period varies in intensity in different animals.

There was a well-known elephant some years ago in the Balaghat district of the Central Provinces which became historical through the extraordinary malignity of its disposition. Having escaped from the fetters, it killed the mahout, and at once made off towards the forests. It is a curious example of nature that creatures (ferae naturae) have a tendency to return to their original state of savagedom when the opportunity is offered. If an elephant is seized with a panic when upon open ground, it will rush for the nearest jungle, probably with the intention of concealment. The animal in question returned to its wild state directly it had escaped from confinement, but the domestication of many years appears to have sharpened its intellect, and to have exaggerated its powers for mischief and cunning. It became the scourge, not only of the immediate neighbourhood, but of a considerable portion of a district which included an area of a hundred miles in length by forty or fifty in width.

No village was safe from the attack of this infuriated beast. It would travel great distances, and appear at unexpected intervals, suddenly presenting itself to the horrified villagers, who fled in all directions, leaving their homes and their supplies of grain to be demolished by the omnipotent intruder, who tore down their dwellings, ransacked their stores of corn, and killed any unfortunate person who came within its reach.

There was a cruel love of homicide in this animal that has rarely been recorded. Not only would it attack villages in pursuit of forage, but it was particularly addicted to the destruction of the lofty watching-places in the fields, occupied nightly by the villagers to scare wild animals from their crops. These watch-houses are generally constructed upon strong poles secured by cross-pieces, on the top of which, about sixteen feet from the ground, is a small hut upon a platform. This is thatched to protect the occupant from the heavy dew or rain. From such elevated posts the watchers yell and scream throughout the night to frighten the wild beasts. To attack and tear down such posts was the delight of this bloodthirsty elephant. Instead of being scared by the shouts of the inmates, it was attracted by their cries, and, unseen in the dark, it was upon them almost before they were aware of its presence. The strong posts upon which the constructions had been raised offered no resistance to the attack, and the miserable watchers found themselves hurled to the ground together with the ruins of their upturned shelter. In another moment they were either caught and stamped to death, or chased through the darkness by the pursuing elephant, and when captured they were torn limb from limb, as the brute exhibited a cruel satisfaction in placing one foot upon the victim, and then tearing with its trunk an arm, a leg, or the head from the mangled body.

In this manner the elephant killed upwards of twenty people throughout the district, and it became absolutely necessary, if possible, to destroy it.

This was at last effected by Colonel Bloomfield and a friend, who determined at all hazards to hunt it down by following through the jungles, guided by the reports of the natives, who were on the lookout in all directions. The animal showed peculiar cunning, as it never remained in the same place, but travelled a considerable distance immediately after the committal of some atrocity, and concealed itself within the jungles until prompted to another raid in some new direction. I am indebted to Colonel Bloomfield for an interesting description of the manner in which, after many days of great fatigue and patience, he at length succeeded, with the assistance of native trackers, in discovering this formidable opponent, asleep within a dense mass of thorns and grass in the heart of an extensive jungle. The elephant awoke before they could distinctly see its form, owing to the extreme thickness of the covert, but the fight commenced. There was a considerable difference between the attack upon defenceless villagers, who fled before it in hopeless panic, and a stand-up fight with two experienced European shikaris armed with the best rifles; the terror of the district quickly showed its appreciation of discretion, and, badly wounded, it retreated through the forest, well followed by the determined hunters. Again and again it was overtaken, and a shot was taken whenever the dense jungle afforded an opportunity. At length, maddened by pursuit and wounds, it turned to charge, thereby exposing itself in an open place, and both bullets crashed into its brain, the shot from Colonel Bloomfield's rifle passing completely through its head.

It would be impossible to determine whether such an elephant could have been subdued and re-domesticated had its capture been effected. There are many cases on record where a "must" elephant has committed grievous depredations, after killing those who were its ordinary attendants, but when re-captured, the temporary excitement has passed away, and the animal has become as harmless as it was before the period of insanity. Mr. G. P. Sanderson, the superintendent of the Government keddahs in Assam, gives a vivid description of an elephant that escaped after killing its mahout and several villagers in the neighbourhood. This animal, like Colonel Bloomfield's elephant, already described, became the terror of the district, and destroyed many villagers, until it was decided by the authorities to attempt its destruction.

Mr. Sanderson was of opinion that it was too valuable to be heedlessly sacrificed; he therefore determined to capture it alive, if possible, through the aid of certain clever elephants belonging to the keddah establishment.

The police of the district were ordered to obtain the necessary information, and the malefactor was reported after a few days to have destroyed another village, where it remained, devouring the rice and grain in the absence of the panic-stricken villagers.

No time was lost in repairing to the spot with three highly-trained elephants, two of which were females; the third was a well-known fighting male, a tusker named Moota Gutche, who was usually employed to dominate the obstreperous wild elephants when refractory in the keddah enclosures. The necessary ropes and chains were prepared, and the small but experienced party started, Mr. Sanderson being armed only with a long spear, and riding on the pad, well girthed upon the back of Moota Gutche.

A short hour's march brought them in sight of a ruined village on a level plain, which skirted a dense forest. When within a quarter of a mile, a large male elephant was discovered restlessly walking to and fro as though keeping guard over the ruins he had made. This was the culprit taken in the act.

Leaving the two females in the rear, with instructions to follow upon a given signal, Mr. Sanderson on Moota Gutche advanced slowly to the encounter. The rogue elephant did not appear to notice them until within about 200 yards; it then suddenly halted, and turning round, it faced them as though in astonishment at being disturbed. This attitude did not last very long, as Moota Gutche still advanced until within ninety or a hundred paces. The elephants now faced each other, and Moota Gutche began to lower his head when he observed his antagonist backing a few paces, which he well knew was the customary preparation for a charge. "Reculez pour mieux sauter" was well exemplified when in another moment the vagrant elephant dashed forward at great speed to the attack, trumpeting and screaming with mad fury. In the meantime Moota Gutche coolly advanced at a moderate pace. The shock of the encounter was tremendous. The spear flew out of the rider's hands with the collision, but Moota Gutche was a trained fighter, and having lowered his head, which had for the moment exposed his mahout, he quickly caught his opponent under the throat with its neck between his tusks, and then bearing upwards, he forced the head of his adversary high in the air; now driving forwards with all his strength, he hurled the other backwards, and with a dexterous twist he threw it upon its side and pinned it to the ground. In an instant Mr. Sanderson slipped off and secured the hind legs with a strong rope. The two females quickly arrived, and within a few minutes the late terror of the neighbourhood was helplessly fettered, and was led captive between the females towards the camp from which it had escaped, assisted, when obstreperous, by the tusks of Moota Gutche applied behind.

This elephant completely recovered from its temporary madness, and became a useful animal, affording a striking example of the passing insanity of the male passion, and the power of careful management in subduing a brute of such stupendous force.

After this incident Moota Gutche, with about forty of the keddah elephants, was kindly lent to me by Mr. Sanderson during a shooting excursion of twenty-five days upon the "churs" or islands of the Brahmaputra river south of Dhubri. In India the tiger is so commonly associated with the elephant that in describing one it is impossible to avoid a connection with the other.

Moota Gutche was a peculiar character, not altogether amiable, but it was as well to have him upon your own side. During the trip my friend Sanderson was ill with fever, and could not accompany me. I was therefore at the disadvantage of being the only gun in a long line of elephants, which would on ordinary occasions have been manned by at least four guns. At first I imagined that my trip would be a failure, as I knew a mere nothing of the language, and the elephants and their mahouts were alike strangers to me, but I soon discovered that their excellent training as keddah servants constantly employed in the capture of wild elephants under their indefatigable superintendent, Mr. Sanderson, rendered them capable almost instinctively of understanding all my ways, and we became excellent friends, both man and beast.

I arranged my long line of elephants according to their paces and dispositions, and each day they preserved the same positions, so that every mahout knew his place, and the elephants were accustomed to the animals upon the right and left. In the centre were the slowest, and upon either flank were the fastest elephants, while two exceedingly speedy animals, with intelligent mahouts, invariably acted as scouts, generally a quarter of a mile ahead on either flank.

My own elephant was accompanied on one side by Moota Gutche, on the other by a rough but dependable character whose name I have forgotten. I kept these always with me, as they were useful in the event of a tiger that would not bolt from the dense wild-rose thickets, in which case our three elephants could push him out.

This arrangement was perfect, and after a few days' experience our line worked with the precision of well-drilled cavalry; sometimes, with extra elephants, I had as many as fifty in the field. The result of this discipline was that no tiger or leopard ever escaped if once on foot; although hunted in some instances for hours, the animal was invariably killed. A remarkable instance of this occurred at the large island of Bargh Chur, which includes several thousand acres, the greater portion being covered with enormous grass and dense thickets of tamarisk, which, in the hot season, is the cool and loved resort of tigers. There were also extensive jungles in swampy portions of the island, so intermixed with reeds and marsh grass of twelve or fourteen feet high, that it was difficult to penetrate, even upon an elephant.

I was out at the usual early hour, shortly after sunrise, the shikaris having returned to camp with the news that none of the bullocks tied up for baits during the preceding night had been killed; it therefore remained to try our fortune by simply beating the high grass jungle in line, on speculation, and in the same manner to drive the occasional dense coverts of feathery tamarisk.

We had proceeded with a line of about five-and-thirty elephants, well extended ten yards apart, and in this manner we had advanced about a mile, when our attention was attracted by a native calling to us from a large ant-hill which enabled him to be distinguished above the grass. We immediately rode towards him, and were informed that a tiger had killed his cow the night before, and had dragged the body into jungle so dense that he had been afraid to follow. This was good news; we therefore took the man upon an elephant as our guide towards the reported spot.

The elephants continued to advance in line, occasionally disturbing wild pigs and hog deer, which existed in great numbers, but could hardly have been shot even had I wished, as the grass was so thick and long that the animals could not be seen; there were only signs of their disturbance by the sudden rush and the waving of the grass just in front of the advancing elephants, who were thus kept in continual excitement.

In about twenty minutes we emerged from the high grass upon a great extent of highly cultivated land, where the sandy loam had been reduced to the fine surface of a well-kept garden. Bordering upon this open country was an extensive jungle composed of trees averaging about a foot in diameter, but completely wedged together among impenetrable reeds fully eighteen feet in length, and nearly an inch in thickness, in addition to a network of various tough creepers, resulting from a rich soil that was a morass during the rainy season. Although the reeds appeared tolerably dry, they would not burn, as there were signs among some half-scorched places where attempts had been recently made to fire the jungle.

Our guide soon pointed to the spot where his cow had been dragged by the tiger into this formidable covert. There was no mistake about the marks, and the immense tracks in the soft ground proved the size and sex of the destroyer.

Nobody questioned the fact of the tiger being at home, and the only question was "how to beat him out." The jungle was quite a mile in length without a break in its terrible density; it was about half a mile in width, bounded upon one side by the cleared level ground in cultivation, and on the other by the high grass jungle we had left, but this had been partially scorched along the edge in the attempts to burn.

A good look-out would have spied any animal at a hundred and fifty yards had it attempted to leave the jungle.

As the country was a dead level, it was difficult to forecast the retreat of a tiger when driven from such a thicket, and it was a serious question whether it would be possible to dislodge him.

Whenever you commence a drive, the first consideration should be, "If the animal is there, where did it come from?"—-as it will in all probability attempt to retreat to that same locality. There was no possibility of guessing the truth in such a country of dense grass, and with numerous islands of the same character throughout this portion of the Brahmaputra, but there was one advantage in the fact that one side was secure, as the tiger would never break covert upon the cultivated land; there remained the opposite side, which would require strict watching, as he would probably endeavour to slink away through the high grass to some distant and favourite retreat.

I therefore determined to take my stand at the end of the thick jungle which we had passed upon arrival, at the corner where it joined the parched grass that had been fire-scorched, and near the spot where the cow had been dragged in. I accordingly sent the elephants round to commence the drive about two hundred yards distant, entering from the cultivated side and driving towards me, as I concluded the tiger in such massive jungle would not be far from the dead body. At the same time, I sent two scouting elephants to occupy positions outside the jungle on the high grass side, within sight of myself; I being posted on my elephant at the corner, so that I commanded two views — -the end, and the grass side.

My signal, a loud whistle, having been given, the line of elephants advanced towards my position. The crashing of so many huge beasts through the dense crisp herbage sounded in the distance like a strong wind, varied now and then by the tearing crunch as some opposing branches were torn down to clear the way.

I was mounted upon a female elephant, a good creature named Nielmonne, who was reputed to be staunch, but as the line of beaters approached nearer, and the varied sounds increased in intensity, she became very nervous and restless, starting should a small deer dart out of the jungle, and evidently expecting momentarily the appearance of the enemy. There are very few elephants that will remain unmoved when awaiting the advance of a line of beaters, whether they may be of their own species or human beings. On this occasion the rushing sound of the yielding jungle, which was so thick as to test the elephants' powers in clearing a passage through it, was presently varied by a sharp trumpet, then by a low growl, followed by that peculiar noise emitted by elephants when excited, resembling blows upon a tambourine or kettle-drum. This is a sound that invariably is heard whenever an elephant detects the fresh scent of a tiger; and Nielmonne, instead of standing quiet, became doubly excited, as she evidently understood that the dreaded game was on foot, and advancing before the line.

As I was posted at the sharp angle of the corner, I presently observed several elephants emerge upon my left and right, as the line advanced with wonderful regularity, and so close were the animals together that it was most unlikely any tiger could have broken back.

My servant Michael was behind me in the howdah. He was a quiet man, who thoroughly understood his work, and seldom spoke without being first addressed. On this occasion he broke through the rule. "Nothing in this beat, sahib," he exclaimed . . . . "Hold your tongue, Michael, till the cover's beaten out. Haven't I often told you that you can't tell what's in the jungle until the last corner is gone through?"

Nearly all the elephants were now out, and only about half a dozen remained in the jungle, all still advancing in correct line, and perhaps a dozen yards remaining of dense reeds and creepers forming the acute angle at the extremity. They still came on. Two or three of the mahouts shouted, "The tiger's behind, we must go back and take a longer beat." Nothing remained now except six or seven yards of the sharp corner, and the elephants marched forward, when a tremendous roar suddenly startled them in all directions, and one of the largest tigers I have ever seen sprang forward directly towards Nielmonne, who, I am ashamed to say, spun round as though upon a pivot, and prevented me from taking a most splendid shot. The next instant the tiger had bounded back with several fierce roars, sending the line of elephants flying, and once more securing safety in the almost impervious jungle from which he had been driven.

This was a most successful drive, but a terrible failure, owing entirely to the nervousness of my elephant. I never saw a worse jungle, and now that the tiger had been moved, it would be doubly awkward to deal with him, as he would either turn vicious and spring upon an elephant unawares from so dense a covert, or slink from place to place as the line advanced, but would never again face the open.

I looked at my watch; it was exactly half-past eight. The mahouts suggested that we should not disturb him, but give him time to sleep, and then beat for him in the afternoon. I did not believe in sleep after he had been so rudely aroused by a long line of elephants, but I clearly perceived that the mahouts did not enjoy the fun of beating in such dreadful jungle, and this they presently confessed, and expressed a wish to have me in the centre of the line, as there was no gun with the elephants should the tiger attack.

I knew that I should be useless, as it would be impossible to see a foot ahead in such dense bush, but to give them confidence I put my elephant in line, and sent forward several scouting elephants to form a line along a narrow footpath which cut the jungle at right angles about a quarter of a mile distant.

Once more the line advanced, the elephants marching shoulder to shoulder, and thus bearing down everything before them, as I determined to take the jungle backwards and forwards in this close order lest the wary tiger might crouch, and escape by lying close.

Several times the elephants sounded, and we knew that he must be close at hand, but it was absolutely impossible to see anything beyond the thick reedy mass, through which the line of elephants bored as through a solid obstacle.

Three times with the greatest patience we worked the jungle in this searching manner, when on the third advance I left the line, finding the impossibility of seeing anything, and took up my position outside the jungle on the cultivated land, exactly where the footpath was occupied by the scout elephants at intervals, which intersected the line of advance.

Presently there was a commotion among the elephants, two or three shrill trumpets, then the kettle-drum, and for a moment I caught sight of a dim shadowy figure stealing through some high reeds upon the border which fringed the jungle. I immediately fired, although the elephant was so unsteady that I could not be sure of the shot; also the object was so indistinct, being concealed in the high reeds, that I should not have observed it upon any other occasion than our rigid search. Immediately afterwards, a shout from one of the mahouts upon a scouting elephant informed us that the tiger had crossed the path and had gone forward, having thus escaped from the beat!

Here was fresh work cut out! Up to this moment we had managed to keep him within an area of a quarter of a mile in length, by half a mile in width; he had now got into new ground, and was in about a three-quarter mile length of the same unbeaten jungle.

There was nothing else to do but to pursue the same tactics, and we patiently continued to beat forward and backward, again and again, but without once sighting our lost game. It was half-past twelve, and the sun was burning hot, the sky being cloudless. The elephants once more emerged from the sultry jungle; they were blowing spray with their trunks upon their flanks, from water sucked up from their stomachs; and the mahouts were all down-hearted and in despair. "It's of no use," they said, "he's gone straight away, who can tell where? When you fired, perhaps you wounded him, or you missed him; at any rate, he's frightened and gone clean off, we shall never see him again; the elephants are all tired with the extreme heat, and we had better go to the river for a bath."

I held a council of war, with the elephants in a circle around me. It is of no use to oppose men when they are disgusted, you must always start a new idea. I agreed with my men, but I suggested that as we were all hot, and the elephants fatigued, the tiger must be in much the same state, as we had kept him on the run since eight o'clock in the morning, I having actually timed the hour "half-past eight" when he charged out of the last corner. "Now," said I, "do you remember that yesterday evening I killed a buck near some water in a narrow depression in the middle of tamarisk jungle? I believe that is only a continuation of this horrible thicket, and if the tiger is near-ly played out, he would naturally make for the water and the cool tamarisk. You form in line in the jungle here, and give me a quarter of an hour's start, while I go ahead and take up my position by that piece of water. You then come on, and if the tiger is in the jungle, he will come forward towards the water, where I shall meet him; if he's not there, we shall anyhow be on our direct route, and close to our camp by the river."

This was immediately accepted, and leaving the elephants to form line, I hurried forward on Nielmonne, keeping in the grass outside the edge of the long jungle.

I had advanced about three-quarters of a mile, when the character of the jungle changed to tamarisk, and I felt certain that I was near the spot of yesterday. I accordingly ordered the mahout to turn into the thick feathery foliage to the left, in search of the remembered water. There was a slight descent to a long but narrow hollow about 50 or 60 yards wide; this was filled with clear water for an unknown length.

I was just about to make a remark, when, instead of speaking, I gently grasped the mahout by the head as I leaned over the howdah, and by this signal stopped the elephant.

There was a lovely sight, which cheered my heart with that inexpressible feeling of delight which is the reward for patience and hard work. About 120 yards distant on my left, the head and neck of a large tiger, clean and beautiful, reposed above the surface, while the body was cooling, concealed from view. Here was our friend enjoying his quiet bath, while we had been pounding away up and down the jungles which he had left.

The mahout, although an excellent man, was much excited. "Fire at him," he whispered.

"It is too far to make certain," I replied in the same undertone.

"Your rifle will not miss him; fire, or you will lose him. He will see us to a certainty and be off. If so, we shall never see him again," continued Fazil, the mahout.

"Hold your tongue," I whispered. "He can't see us, the sun is at our back, and is shining in his eyes —- see how green they are."

At this moment of suspense the tiger quietly rose from his bath, and sat up on end like a dog. I never saw such a sight. His head was beautiful, and the eyes shone like two green electric lights, as the sun's rays reflected from them, but his huge body was dripping with muddy water, as he had been reclining upon the alluvial bottom.

"Now's the time," whispered the over-eager mahout. "You can kill him to a certainty. Fire, or he'll be gone in another moment."

"Keep quiet, you fool, and don't move till I tell you." For quite a minute the tiger sat up in the same position; at last, as though satisfied that he was in safety and seclusion, he once more lay down with only the head and neck exposed above the surface.

"Back the elephant gently, but do not turn round," I whispered. Immediately Nielmonne backed through the feathery tamarisk without the slightest sound, and we found ourselves outside the jungle. We could breathe freely.

"Go on now, quite gently, till I press your head; then turn to the right, descending through the tamarisk, till I again touch your puggery" (turban).

I counted the elephant's paces as she moved softly parallel with the jungle, until I felt sure of my distance. A slight pressure upon the mahout's head, and Nielmonne turned to the right. The waving plumes of the dark-green tamarisk divided as we gently moved forward, and in another moment we stopped. There was the tiger in the same position, exactly facing me, but now about 75 paces distant.

"Keep the elephant quite steady," I whispered; and, sitting down upon the howdah seat, I took a rest with the rifle upon the front bar of the gun-rack. A piece of tamarisk kept waving in the wind just in front of the rifle, beyond my reach. The mahout leaned forward and gently bent it down. Now, all was clear. The tiger's eyes were like green glass. The elephant for a moment stood like stone. I touched the trigger.

There was no response to the loud report of 6 drams of powder from the '577 rifle, no splash in the unbroken surface of the water. The tiger's head was still there, but in a different attitude, one-half below the surface, and only one cheek, and one large eye still glittering like an emerald, above.

"Run in quick,"—-and the order was instantly obeyed, as Nielmonne splashed through the pool towards the silent body of the tiger. There was not a movement of a muscle. I whistled loud, then looked at my watch—-on the stroke of 1 P.M. From 8.30 till that hour we had worked up that tiger, and although there was no stirring incident connected with him, I felt very satisfied with the result.

In a short time the elephants arrived, having heard the shot, followed by my well-known whistle. Moota Gutche was the first to approach; and upon observing the large bright eye of the tiger above water, he concluded that it was still alive; he accordingly made a desperate charge, and taking the body on his tusks, he sent it flying some yards ahead; not content with this display of triumph, he followed it up, and gave it a football-kick that lifted it clean out of the water. This would have quickly ended in a war-dance upon the prostrate body, that would have crushed it and destroyed the skin, had not the mahout, with the iron driving-hook, bestowed some warning taps upon the crown of Moota Gutche's head that recalled him to a calmer frame of mind. A rope was soon made fast to the tiger's neck, and Moota Gutche hauled it upon dry ground, where it was washed as well as possible, and well scrutinized for a bullet-hole.

There was no hole whatever in that tiger. The bullet having entered the nostril, broken the neck, and run along the body, the animal consequently had never moved. The first shot, when obscured in thick jungle, had probably deflected from the interposing reeds—-at all events it missed. This tiger, when laid out straight, but without being pulled to increase its length, measured exactly 9 feet 8 inches from nose to tail.

CHAPTER III

THE ELEPHANT (continued)

The foregoing chapter is sufficient to explain the ferocity of the male elephant at certain seasons which periodically affect the nervous system. It would be easy to multiply examples of this cerebral excitement, but such repetitions are unnecessary. The fact remains that the sexes differ materially in character, and that for general purposes the female is preferred in a domesticated state, although the male tusker is far more powerful, and when thoroughly trustworthy is capable of self-defence against attack, and of energy in work that would render it superior to the gentler but inferior female. (The female differs from other quadrupeds in the position of her teats, which are situated upon the breast between the fore legs. She is in the habit of caressing her calf with her trunk during the operation of suckling.)

It may be inferred that a grand specimen of a male elephant is of rare occurrence. A creature that combines perfection of form with a firm but amiable disposition, and is free from the timidity which unfortunately distinguishes the race, may be quite invaluable to any resident in India. The actual monetary value of an elephant must of necessity be impossible to decide, as it must depend upon the requirements of the purchaser and the depth of his pocket. Elephants differ in price as much as horses, and the princes of India exhibit profuse liberality in paying large sums for animals that approach their standard of perfection.

The handsomest elephant that I have ever seen in India belongs to the Rajah of Nandgaon, in the district bordering upon Reipore. I saw this splendid specimen among twenty others at the Durbar of the Chief Commissioner of the Central Provinces in December 1887, and it completely eclipsed all others both in size and perfection of points. The word "points" is inappropriate when applied to the distinguishing features of an elephant, as anything approaching the angular would be considered a blemish. An Indian elephant to be perfect should be 9 feet 6 inches in perpendicular height at the shoulder. The head should be majestic in general character, as large as possible, — especially broad across the forehead, and well rounded. The boss or prominence above the trunk should be solid and decided, mottled with flesh-coloured spots; these ought to continue upon the cheeks, and for about three feet down the trunk. This should be immensely massive; and when the elephant stands at ease, the trunk ought to touch the ground when the tip is slightly curled. The skin of the face should be soft to the touch, and there must be no indentations or bony hollows, which are generally the sign of age. The ears should be large, the edges free from inequalities or rents, and above all they ought to be smooth, as though they had been carefully ironed. When an elephant is old, the top of the ear curls, and this symptom increases with advancing years. The eyes should be large and clear, the favourite colour a bright hazel. The tusks ought to be as thick as possible, free from cracks, gracefully curved, very slightly to the right and left, and projecting not less than three feet from the lips. The body should be well rounded, without a sign of any rib. The shoulders must be massive with projecting muscular development; the back very slightly arched, and not sloping too suddenly towards the tail, which should be set up tolerably high. This ought to be thick and long, the end well furnished with a double fringe of very long thick hairs or whalebone-looking bristles. The legs should be short in proportion to the height of the animal, but immensely thick, and the upper- portion above the knee ought to exhibit enormous muscle. The knees should be well rounded, and the feet be exactly equal to half the perpendicular height of the elephant when measured in their circumference, the weight pressing upon them whilst standing.

The skin generally ought to be soft and pliable, by no means tight or strained, but lying easily upon the limbs and body.

An elephant which possesses this physical development should be equal in the various points of character that are necessary to a highly-trained animal.

When ordered to kneel, it should obey instantly, and remain patiently upon the ground until permitted to rise from this uneasy posture. In reality the elephant does not actually kneel upon its fore knees, but only upon those of the hinder legs, while it pushes its fore legs forward and rests its tusks upon the ground. This is a most unnatural position, and is exceedingly irksome. Some elephants are very impatient, and they will rise suddenly without orders while the ladder is placed against their side for mounting. Upon one occasion a badly-trained animal jumped up so suddenly that Lady Baker, who had already mounted, was thrown off on one side, while I, who was just on the top of the ladder, was thrown down violently upon the other. A badly-tutored elephant is exceedingly dangerous, as such vagaries are upon so large a scale that a fall is serious, especially should the ground be stony.

A calm and placid nature free from all timidity is essential. Elephants are apt to take sudden fright at peculiar sounds and sights. In travelling through a jungle path it is impossible to foretell what animals may be encountered on the route. Some elephants will turn suddenly round and bolt, upon the unexpected crash of a wild animal startled in the forest. The scent or, still worse, the roar of a bear within 50 yards of the road will scare some elephants to an extent that will make them most difficult of control. The danger may be imagined should an elephant absolutely run away with his rider in a dense forest; if the unfortunate person should be in a howdah he would probably be swept off and killed by the intervening branches, or torn to shreds by the tangled thorns, many of which are armed with steel-like hooks.

It is impossible to train all elephants alike, and very few can be rendered thoroughly trustworthy; the character must be born in them if they are to approach perfection.

Our present perfect example should be quite impassive, and should take no apparent notice of anything, but obey his mahout with the regularity of a machine. No noise should disturb the nerves, no sight terrify, no attack for one moment shake the courage; even the crackling of fire should be unheeded, although the sound of high grass blazing and exploding before the advancing line of fire tries the nerves of elephants more than any other danger.

An elephant should march with an easy swinging pace at the rate of 5 miles an hour, or even 6 miles within that time upon a good flat road. As a rule, the females have an easier pace than the large males. When the order to stop is given, instead of hesitating, the elephant should instantly obey, remaining rigidly still without swinging the head or flapping the ears, which is its inveterate and annoying habit. The well-trained animal should then move backward or forward, either one or several paces, at a sign from the mahout, and then at once become as rigid as a rock.

Should the elephant be near a tiger, it will generally know the position of the enemy by its keen sense of smell. If the tiger should suddenly charge from some dense covert with the usual short but loud roars, the elephant ought to remain absolutely still to receive the onset, and to permit a steady aim from the person in the howdah. This is a very rare qualification, but most necessary in a good shikar elephant. Some tuskers will attack the tiger, which is nearly as bad a fault as running in the opposite direction; but the generality, even if tolerably steady, will swing suddenly upon one side, and thus interrupt the steadiness of aim.

The elephant should never exercise its own will, but ought to wait in all cases for the instructions of the mahout, and then obey immediately.

Such an animal, combining the proportions and the qualities I have described, might be worth in India about / 1500 to any Indian Rajah, but there may be some great native sportsmen who would give double that amount for such an example of perfection, which would combine the beauty required for a state elephant, with the high character of a shikar animal.

Native princes and rajahs take a great pride in the trappings of their state elephants, which is exhibited whenever any pageant demands an extraordinary display. I have seen cloths of silk so closely embroidered with heavy gold as to be of enormous value, and so great a weight that two men could barely lift them. Such cloths may have been handed down from several generations, as they are seldom used excepting in the state ceremonies which occur at distant intervals. A high caste male elephant in its gold trappings, with head-piece and forehead lap equally embroidered, and large silver bells suspended from its tusks, is a magnificent object during the display attending a durbar. At such an occasion there may be a hundred elephants all in their finery, each differing from the other both in size and in the colours of their surroundings.

The outfit for an elephant depends upon the work required. The first consideration is the protection of the back. Although the skin appears as though it could resist all friction, it is astonishing how quickly a sore becomes established, and how difficult this is to heal. The mahouts are exceedingly careless, and require much supervision; the only method to ensure attention is to hold them responsible and to deduct so many rupees from their pay should the backs of their animals be unsound.

With proper care an elephant ought never to suffer, as the pad should be made to fit its figure specially. The usual method is to cover the back from the shoulders to the hips with a large quilted pad stuffed with cotton, about 2 1/2 inches thick. In my opinion, wool is preferable to cotton, and, instead of this coverlet being compact, there should be an opening down the centre, to avoid all pressure upon the spine. A quilted pad stuffed with wool, 3 inches thick, with an opening down the middle, would rest comfortably upon the animal's back, and would entirely relieve the highly-arched backbone, which would thus be exposed to a free current of air, and would remain hard instead of becoming sodden through perspiration. Upon this soft layer the large pad is fixed. This is made of the strongest sacking, stuffed as tight as possible with dried reeds of a tough variety that is common in most tanks; this is open in the centre and quite a foot thick at the sides, so that it fills up the hollow, and rests the weight upon the ribs at a safe distance from the spine.

There are various contrivances in the shape of saddles. The ordinary form for travelling is the char-jarma; this is an oblong frame, exceedingly strong, which is lashed upon the pad secured by girths. It is stuffed with cotton, and neatly covered with native cloth. A stuffed back passes down the centre like a sofa, and two people on either side sit dos-a-dos, as though in an Irish car. Iron rails protect the ends, and swing foot-boards support the feet. This is, in my opinion, the most comfortable way of riding, but some care is necessary in proportioning the weights to ensure a tolerable equilibrium, otherwise, should the route be up and down steep nullahs, the char-jarma will shift upon one side, and become most disagreeable to those who find themselves on the lower level. Natives prefer a well-stuffed pad, as they are accustomed to sit with their legs doubled up in a manner that would be highly uncomfortable to Europeans. Such pads are frequently covered with scarlet cloth and gold embroidery, while the elephant is dressed in a silk and gold cloth reaching to its knees. The face and head are painted in various colours and devices, exhibiting great taste and skill on the part of the designer. It is curious to observe the dexterity with which an otherwise ignorant mahout will decorate the head of his animal by drawing most elaborate curves and patterns, that would tax the ability of a professional artist among Europeans.

The howdah is the only accepted arrangement for sporting purposes, and much attention is necessary in its construction, as the greatest strength should be combined with lightness. There ought to be no doors, as they weaken the solidity of the whole. The weight of a good roomy howdah should not exceed two hundredweight, or at the outside 230 pounds. It must be remembered that the howdah is not adapted for travelling, as there is a disagreeable swinging motion inseparable from its position upon the elephant's back which is not felt upon either the pad or the char-jarma. The howdah is simply for shooting, as you can fire in any direction, which is impossible from any other contrivance where the rider sits in a constrained position.

A good howdah should be made of exceedingly strong and tough wood for the framework, dovetailed, and screwed together, the joints being specially secured by long corner straps of the best iron. The frame ought to be panelled with galvanised wire of the strongest description, the mesh being one-half inch. The top rail, of a hard wood, should be strengthened all around the howdah by the addition of a male bamboo 1 1/2 inch in diameter, securely lashed with raw hide, so as to bind the structure firmly together, and to afford a good grip for the hand. As the howdah is divided into two compartments, the front being for the shooter, and the back part for his servant, the division should be arranged to give increased strength to the construction by the firmness of the cross pieces, which ought to bind the sides together in forming the middle seat; the back support of which should be a padded shield of thick leather, about 15 inches in diameter, secured by a broad strap of the same material to buckles upon the sides. This will give a yielding support to the back of the occupant when sitting. The seat should lift up, and be fitted as a locker to contain anything required; and a well-stuffed leather cushion is indispensable. The gun-rack should be carefully arranged to contain two guns upon the left, and one upon the right of the sitter. These must be well and softly padded, to prevent friction. The floor should be covered either with thick cork or cork-matting to prevent the feet from slipping.

It must be remembered that a howdah may be subjected to the most severe strain, especially should a tiger spring upon the head of an elephant, and the animal exert its prodigious strength to throw off its assailant. The irons for fastening the girths should therefore be of the toughest quality, and, instead of actual girths, only thick ropes of cotton ought to be used. A girth secured with a buckle is most dangerous, as, should the buckle give way, an accident of the most alarming kind must assuredly occur. The howdah ought to be lashed upon the elephant by six folds of the strong cotton rope described, tightened most carefully before starting. It should be borne in mind that much personal attention is necessary during this operation, as the natives are most careless. Two or three men ought to sit in the howdah during the process of lacing, so as to press it down tightly upon the pad, otherwise it will become loose during the march, and probably lean over to one side, which is uncomfortable to both man and beast. A large hide of the sambur deer, well cured and greased so as to be soft and pliable, should, invariably protect the belly of the elephant, and the flanks under the fore legs, from the friction of the girthing rope. The breastplate and crupper also require attention. These ought to be of the same quality of cotton rope as used for the girths, but that portion of the crupper which passes beneath the tail should pass through an iron tube bent specially to fit, like the letter V elongated, U. This is a great safeguard against galling, and I believe it was first suggested by Mr. G. P. Sanderson.

A fine male elephant, well accoutred with his howdah thoroughly secured, and a good mahout, is a splendid mount, and the rider has the satisfaction of feeling that his animal is well up to his weight. I do not know a more agreeable sensation than the start in the early morning upon a thoroughly dependable elephant, with all the belongings in first-rate order, and a mahout who takes a real interest in his work; a thorough harmony exists between men and beast, the rifles are in their places, and you feel prepared for anything that may happen during the hazardous adventures of the day.

But how much depends upon that mahout! It is impossible for an ordinary bystander to comprehend the secret signs which are mutually understood by the elephant and his guide, the gentle pressure of one toe, or the compression of one knee, or the delicate touch of a heel, or the almost imperceptible swaying of the body to one side; the elephant detects every movement, howsoever slight, and it is thus mysteriously guided by its intelligence; the mighty beast obeys the unseen helm of thought, just as a huge ship yields by apparent instinct to the insignificant appendage which directs her course, the rudder. All good riders know the mystery of a "good hand" upon a horse; this is a thing that is understood, but cannot be described except by a negative. There are persons who can sit a horse gracefully and well, but who have not the instinctive gift of hand. The horse is aware of this almost as soon as the rider has been seated in the saddle. In that case, whether the horse be first-class or not, there will be no comfort for the animal, and no ease for the rider.

If such a person puts his horse at a fence, the animal will not be thoroughly convinced that his rider wishes him to take it. There are more accidents occasioned by a "bad hand" than by any other cause. If this is the case with a horse well bitted, what must be the result should an elephant be guided by a mahout of uncertain temperament? The great trouble when travelling on an elephant is the difficulty in getting the mahout to obey an order immediately, and at the same time to convey that order to the animal without the slightest hesitation. Natives frequently hesitate before they determine the right from left. This is exasperating to the highest degree, and is destructive to the discipline of an elephant. There must be no uncertainty; if there is the slightest vacillation, it will be felt instinctively in the muscles of the rider, and the animal, instead of obeying mechanically the requisite pressure of knee or foot, feels that the mahout does not exactly know what he is about. This will cause the elephant to swing his head, instead of keeping steady and obeying the order without delay. In the same manner, when tiger-shooting, the elephant will at once detect anything like tremor on the part of his mahout. Frequently a good elephant may be disgraced by the nervousness of his guide, nothing being so contagious as fear.

Although I may be an exception in the non-admiration of the elephant's sagacity to the degree in which it is usually accepted, there is no one who more admires or is so foolishly fond of elephants. I have killed some hundreds in my early life, but I have learnt to regret the past, and 1 nothing would now induce me to shoot an elephant unless it were either a notorious malefactor, or in self-defence. There is, however, a peculiar contradiction in the character of elephants that tends to increase the interest in the animal. If they were all the same, there would be a monotony; but this is never the case, either among animals or human beings, although they may belong to one family. The elephant, on the other hand, stands so entirely apart from all other animals, and its performances appear so extraordinary owing to the enormous effect which its great strength produces instantaneously, that its peculiarities interest mankind more than any smaller animal. Yet, when we consider the actual aptitude for learning, or the natural habits of the creature, we are obliged to confess that in proportion to its size the elephant is a mere fool in comparison with the intelligence of many insects. If the elephant could form a home like the bee, and store up fodder for a barren season; if it could build a nest of comfort like a bird, to shelter itself from inclement weather; if it could dam up a river like the beaver, to store water for the annual drought; if it could only, like the ordinary squirrel or field mouse, make a store for a season of scarcity, how marvellous we should think this creature, simply because it is so huge! It actually does nothing remarkable, unless specially instructed; but it is this inertia that renders it so valuable to man. If the elephant were to be continually exerting its natural intelligence, and volunteering all manner of gigantic performances in the hope that they would be appreciated by its rider, it would be unbearable; the value of the animal consists in its capacity to learn, and in its passive demeanour, until directed by the mahout's commands.

Nothing can positively determine the character of any elephant; every animal, I believe, varies more or less in courage according to its state of health, which must influence the nervous system. The most courageous man may, if weakened by sickness, be disgusted with himself by starting at an unexpected sound, although upon ordinary occasions he would not be affected. Animals cannot describe their feelings, and they may sometimes feel "out of sorts" without being actually ill, but the nervous system may be unstrung.

I once saw a ridiculous example of sudden panic in an otherwise most dependable elephant. This was a large male belonging to the Government, which had been lent to me for a few months, and was thoroughly staunch when opposed to a charging tiger; in fact, I believe that Moolah Bux was afraid of nothing, and he was the best shikar elephant I have ever ridden. One day we were driving a rocky hill for a tiger that was supposed to be concealed somewhere among the high grass and broken boulders, and, as the line of beaters was advancing, I backed the elephant into some thick jungle, which commanded an open but narrow glade at the foot of the low hill. Only the face of the elephant was exposed, and as this was grayish brown, something similar to the colour of the leafless bushes, we were hardly noticeable to anything that might break covert.

The elephant thoroughly understood the work in hand; and as the loud yells and shouts of the beaters became nearer, Moolah Bux pricked his ears and kept a vigilant look-out. Suddenly a hare emerged about 100 yards distant; without observing our well-concealed position it raced at full speed directly towards us, and in a few seconds it ran almost between the elephant's legs as it made for the protection of the jungle. The mighty Moolah Bux fairly bolted with a sudden terror as this harmless and tiny creature dashed beneath him, and although he recovered himself after 5 or 6 yards, nevertheless for the moment the monster was scared almost by a mouse.

It is this uncertainty of character that has rendered the elephant useless for military purposes in the field since the introduction of fire-arms. In olden times there can be no doubt that a grand array of elephantine cavalry, with towers containing archers on their backs, would have been an important factor when in line of battle; but elephants are useless against fire-arms, and in our early battles with the great hordes brought against us by the princes of India, their elephants invariably turned tail, and added materially to the defeat of their army.

Only a short time ago, at Munich, a serious accident was occasioned by a display of ten or twelve elephants during some provincial fete, when they took fright at the figure of a dragon vomiting fire, and a general stampede was the consequence, resulting in serious injuries to fifteen or sixteen persons.

I once had an elephant who ought to have killed me upon several occasions through sheer panic, which induced him to run away like a railway locomotive rushing through a forest. This was the tusker Lord Mayo, who, although a good-tempered harmless creature, appeared to be utterly devoid of nerves, and would take fright at anything to which it was unaccustomed. The sound of the beaters when yelling and shouting in driving jungle was quite sufficient to start this animal off in a senseless panic, not always for a short distance, as on one occasion it ran at full speed for upwards of a mile through a dense forest, in spite of the driving-hook of the mahout, which had been applied with a maximum severity.

It is curious to observe how all the education of an elephant appears to vanish when once the animal takes fright and bolts for the nearest jungle. That seems to be the one idea which is an instinct of original nature, to retreat into the concealment of a forest.

I was on one occasion mounted upon Lord Mayo in the Balagh district when the beaters were not dependable. A tiger had killed a bullock at the foot of a wooded hill bordered by an open plain. As the beaters had misbehaved upon several occasions by breaking their line, I determined to take command of the beat in person. I therefore formed the line in the open, with every man equidistant, there being about a hundred and twenty villagers. I had placed my shikari with a rifle in a convenient position about 200 yards in advance, upon a mucharn or platform that had been constructed for myself.

Having after some trouble arranged the beaters in a proper line, I gave the order for an advance. In an instant the shouts arose, and three or four tom-toms added to the din.

I was mounted upon Lord Mayo near the centre of the line in the open glade. No sooner had the noise begun, than a violent panic seized this senseless brute, and without the slightest warning it rushed straight ahead for the thick forest at a pace that would nearly equal that of a luggage train. It was in vain that the mahout dug the iron spike into its head and alternately seized its ears by the unsparing hook, away it ran, regardless of all punishment or persuasion, until it reached the jungle, and with a crash we entered in full career!

Fortunately there was no howdah, only a pad well secured by thick ropes. To clutch these tightly, and to dodge the opposing branches by ducking the head, now swinging to the right, then doubling down upon the left to allow the bending trees to sweep across the pad, then flinging oneself nearly over the flank to escape a bough that threatened instant extermination; all these gymnastics were performed and repeated in a few seconds only, as the panic-stricken brute ploughed its way, regardless of all obstructions, which threatened every instant to sweep us off its back. The active mahout of my other elephant, knowing the character of Lord Mayo, had luckily accompanied us with a spear, and although at the time I was unaware of his presence, he was exerting himself to the utmost in a vain endeavour to overtake our runaway elephant. At first I imagined that the great pace would soon be slackened, and that a couple of hundred yards would exhaust the animal's wind, especially as the ground was slightly rising. Instead of this, it was going like a steam-engine, and if there had been the usual amount of thorny creepers we should have been torn to pieces.

" Keep him straight for the hill," I shouted, as I saw we were approaching an inclination. "Don't let him turn to right or left, keep his head straight for the steep ground;" and the mahout, who had been yelling for assistance, and had lost both his turban and skull-cap, did all that he could by tunnelling into the brute's head with his formidable hook to direct it straight up the hill. I never knew an elephant go at such a pace over rocky ground. Young trees were smashed down, some branches torn, others bent forward, which swung backwards with dangerous force, and yet on we tore without a sign of diminishing speed. How I longed for an anchor to have brought up our runaway ship head to wind! We had the coupling chains upon the pad, and my interpreter, Modar Bux, at length succeeded in releasing these, and in throwing them down for any person following to make use of. After a run of quite half a mile, we fortunately arrived at a really steep portion of the hill, where the rocks were sufficiently large to present a difficulty to any runaway. The mahout who had been following our course, breathless and with bleeding feet, here overtook us. Placing himself in advance of the elephant, who seemed determined to continue its flight among the rocks, he dug the spear deep into the animal's trunk, and kept repeating the apparently cruel thrusts until at length it stopped. Several men now arrived with the coupling chains, which were at length with difficulty adjusted, and the elephant's fore legs were shackled together. It was curious to observe the dexterous manner in which it resisted this operation, and had it not been for the dread of the spear I much doubt whether it could have been accomplished.

This was the first time that I had experienced a runaway elephant, but I soon found that both my steeds were equally untrustworthy. A few weeks after this event we had completed the morning's march and found the camp already prepared for our arrival, at a place called Kassli, which is a central depot for railway sleepers as they are received from the native contractors. These were carefully piled in squares of about twenty each, and covered a considerable area of ground at intervals. A large ox had died that morning, and as it was within 50 yards of the tent it was necessary to remove it; the vultures were already crowded in the surrounding trees waiting for its decomposition. As usual, none of the natives would defile themselves by touching the dead body. I accordingly gave orders that one of the elephants should drag it about a mile down wind away from the camp. Lord Mayo was brought to the spot, and the sweeper, being of a low caste, attached a very thick rope to the hind legs of the ox; the other end being made fast to the elephant's pad in such a manner as to form traces. The elephant did not exhibit the slightest interest in the proceeding, and everything was completed, the body of the ox being about 6 or 7 yards behind.

No sooner did Lord Mayo move forward in obedience to the mahout's command, and feel the tug of the weight attached, than he started off in a panic at a tremendous pace, dragging the body through the lanes between the piles of sleepers, upsetting them, and sending them flying in all directions, as the dead ox caught against the corners; and, helter-skelter, he made for the nearest jungle about 300 yards distant. Fortunately some wood-cutters were there, who yelled and screamed to turn him back; but although this had the effect of driving him from the forest, he now started over the plain down hill, dragging the heavy ox behind as though it had been a rabbit, and going at such a pace that none of the natives could overtake him, although by this time at least twenty men were in full pursuit.

The scene was intensely ridiculous, and the whole village turned out to enjoy the fun of a runaway elephant with a dead ox bounding over the inequalities of the ground; no doubt Lord Mayo imagined that he was being hunted by the carcase which so persistently followed him wherever he went. There was no danger to the driver, as the elephant was kept away from the forest. The ground became exceedingly rough and full of holes from the soakage during the rainy season. This peculiar soil is much disliked by elephants, as the surface is most treacherous, and cavernous hollows caused by subterranean water action render it unsafe for the support of such heavy animals. The resistance of the dead ox, which constantly jammed in the abrupt depressions, began to tell upon the speed, and in a short time the elephant was headed, and surrounded by a mob of villagers. I was determined that he should now be compelled to drag the carcase quietly in order to accustom him to the burden; we therefore attached the coupling chains to his fore legs, and drove him gently, turning him occasionally to enable him to inspect the carcase that had smitten him with panic. In about twenty minutes he became callous, and regarded the dead body with indifference.

Although an elephant is capable of great speed, it cannot jump, neither can it lift all four legs off the ground at the same time; this peculiarity renders it impossible to cross any ditch with hard perpendicular sides that will not crumble or yield to pressure, if such a ditch should be wider than the limit of the animal's extreme pace. If the limit of a pace should be 6 feet, a 7-foot ditch would effectually stop an elephant.

Although the strength of an elephant is prodigious whenever it is fully exerted, it is seldom that the animal can be induced to exhibit the maximum force which it possesses. A rush of a herd of elephants, with a determined will against the enclosure of palisades used for their capture would probably break through the barrier, but they do not appear to know their strength, or to act together.

This want of cohesion is a sufficient proof that in a wild state they are not so sagacious as they have been considered. I do not describe the kraal or keddah, which is so well known by frequent descriptions as the most ancient and practical method of capturing wild elephants; but although in Ceylon the kraal has been used from time immemorial, the Singhalese are certainly behind the age as compared with the great keddah establishments of India. In the latter country there is a ditch inside the palisaded enclosure, which prevents the elephants from exerting their force against the structure; in Ceylon this precaution is neglected, and the elephants have frequently effected a breach in the palisade. In Ceylon all the old elephants captured within the kraal or keddah are considered worthless, and only those of scarcely full growth are valued; in India, all elephants irrespective of their age are valued, and the older animals are as easily domesticated as the young.

The keddah establishment at Dacca is the largest in India, and during the last season, under the superintendence of Mr. G. P. Sanderson, 404 elephants were captured in the Garo Hills, 132 being taken in one drive. It is difficult to believe that any district can continue to produce upon this wholesale scale, and it is probable that after a few years elephants will become scarce in the locality. Nevertheless there is a vast tract of forest extending into Burmah, and the migratory habits of the elephant at certain seasons may continue the supply, especially if certain fruits or foliage attract them to the locality.

This migratory instinct is beyond our powers of explanation in the case of either birds, beasts, or fishes. How they communicate, in order to organise the general departure, must remain a mystery. It is well known that in England, previous to the departure of the swallows, they may be seen sitting in great numbers upon the telegraph wires as though discussing the projected journey; in a few days after, there is not a swallow to be seen.

I once, and only once, had an opportunity of seeing elephants that were either migrating, or had just arrived from a migration. This was between 3 degrees and 4 degrees N. latitude in Africa, between Obbo and Farajok. We were marching through an uninhabited country for about 30 miles, and in the midst of beautiful park-like scenery we came upon the magnificent sight of vast herds of elephants. These were scattered about the country in parties varying in numbers from ten to a hundred, while single bulls dotted the landscape with their majestic forms in all directions. In some places there were herds of twenty or thirty entirely composed of large tuskers; in other spots were parties of females with young ones interspersed, of varying growths, and this grand display of elephantine life continued for at least 2 miles in length as we rode parallel with the groups at about a quarter of a mile distant. It would have been impossible to guess the number, as there was no regularity in their arrangement, neither could I form any idea of the breadth of the area that was occupied. I have often looked back upon that extraordinary scene, and it occurred to me forcibly in after years, when I had 3200 elephants' tusks in one station of Central Africa, which must have represented 1600 animals slain for their fatal ivory.

The day must arrive when ivory will be a production of the past, as it is impossible that the enormous demand can be supplied. I have already explained that the African savage never tames a wild animal, neither does he exhibit any sympathy or pity, his desire being, like the gunner of the nineteenth century, to exterminate. It may be readily imagined that wholesale destruction is the result whenever some favourable opportunity delivers a large herd of elephants into the native hands.

There are various methods employed for trapping, or otherwise destroying. Pitfalls are the most common, as they are simple, and generally fatal. Elephants are thirsty creatures, and when in large herds they make considerable roads in their passage towards a river. They are nearly always to be found upon the same track when nightly approaching the usual spot for drinking or for a bath. It is therefore a simple affair to intercept their route by a series of deep pitfalls dug exactly in the line of their advance. These holes vary in shape; the circular are, I believe, the most effective, as the elephant falls head downwards, but I have seen them made of different shapes and proportions according to the individual opinions of the trappers.

It is exceedingly dangerous, when approaching a river, to march in advance of a party without first sending forward a few natives to examine the route in front. The pits are usually about 12 or 14 feet in depth. These are covered over with light wood, and crossed with slight branches or reeds, upon which is laid some long dry grass; this is covered lightly with soil, upon which some elephant's dung is scattered, as though the animal had dropped it during the action of walking. A little broken grass is carelessly distributed upon the surface, and the illusion is complete. The night arrives, and the unsuspecting elephants, having travelled many miles of thirsty wilderness, hurry down the incline towards the welcome river. Crash goes a leading elephant into a well-concealed pitfall! To the right and left the frightened members of the herd rush at the unlooked-for accident, but there are many other pitfalls cunningly arranged to meet this sudden panic, and several more casualties may arise, which add to the captures on the following morning, when the trappers arrive to examine the position of their pits. The elephants are then attacked with spears while in their helpless position, until they at length succumb through loss of blood.

There is another terrible method of destroying elephants in Central Africa. During the dry season, when the withered herbage from 10 to 14 feet in height is most inflammable, a large herd of elephants may be found in the middle of such high grass that they can only be perceived should a person be looking down from some elevated point. If they should be espied by some native hunter, he would immediately give due notice to the neighbourhood, and in a short time the whole population would assemble for the hunt. This would be arranged by forming a circle of perhaps 2 miles in diameter, and simultaneously firing the grass so as to create a ring of flames around the centre. An elephant is naturally afraid of fire, and it has an instinctive horror of the crackling of flames when the grass has been ignited. As the circle of fire contracts in approaching the encircled herd, they at first attempt retreat until they become assured of their hopeless position; they at length become desperate, being maddened by fear, and panic-stricken by the wild shouts of the thousands who have surrounded them. At length, half-suffocated by the dense smoke, and terrified by the close approach of the roaring flames, the unfortunate animals charge recklessly through the fire, burnt and blinded, to be ruthlessly speared by the bloodthirsty crowd awaiting this last stampede. Sometimes a hundred or more elephants are simultaneously destroyed in this wholesale slaughter. The flesh is then cut into long strips and dried, every portion of the animal being smoked upon frames of green wood, and the harvest of meat is divided among the villages which have contributed to the hunt. The tusks are also shared, a certain portion belonging by right to the various headmen and the chief.

When man determines to commence war with the animal kingdom the result must be disastrous to the beasts, if the human destroyers are in sufficient numbers to ensure success. Although fire-arms may not be employed, the human intelligence must always overpower the brute creation, but man must exist in numerical superiority if the wild beasts are to be fairly vanquished by a forced retreat from the locality. From my own observation I have concluded that wild animals of all kinds will withstand the dangers of traps, pitfalls, fire, and the usual methods for their destruction employed by savages, but they will be rapidly cleared out of an extensive district by the use of fire-arms. There is a peculiar effect in the report of guns which appears to excite the apprehension of danger in the minds of all animals. This is an extraordinary instance of the general intelligence of wild creatures, as they must be accustomed to the reports of thunder since the day of their birth. Nevertheless they draw a special distinction between the loud peal of thunder and the comparatively innocent explosion of a fire-arm.

Many years ago in Ceylon I devoted particular attention to this subject, especially as it affects the character of elephants. How those creatures manage to communicate with each other it is impossible to determine, but the fact remains that a very few days' shooting will clear out an extensive district, although the area may comprise a variation of open prairie with a large amount of forest. I have frequently observed, in the portion of Ceylon known as the Park country, the tracks of elephants in great numbers which have evidently been considerable herds that have joined together in a general retreat from ground which they considered insecure. In that district I have arrived at the proper season, when the grass after burning has grown to the height of about 2 feet, and it has literally been alive with elephants. In a week my late brother General Valentine Baker and myself shot thirty-two, and I sent a messenger to invite a friend to join us, in the expectation of extraordinary sport. Upon his arrival after five or six days, there was not an elephant in the country, excepting two or three old single bulls which always infested certain spots.

The reports of so many heavy rifles, which of necessity were fired every evening at dusk in the days of muzzle-loaders, for the sake of cleaning, must have widely alarmed the country, but independently of this special cause there can be no doubt that after a few days' heavy shooting, the elephants will combine in some mysterious manner and disappear from an extensive district. In many ways these creatures are perplexing to the student of natural history. It would occur to most people that in countries where elephants abound we should frequently meet with those that are sick, or so aged that they cannot accompany the herd. Although for very many years I have hunted both in Asia and Africa I have never seen a sick elephant in a wild state, neither have I ever come across an example of imbecility through age. It is rarely we discover a dead elephant that has not met with a violent death, and only once in my life have I by accident found the remains of a tusker with the large tusks intact. This animal had been killed in a fight, as there were unmistakable signs of a fearful struggle, the ground being trodden deeply in all directions.

It is supposed by the natives that when an elephant is mortally sick it conceals itself in the thickest and most secluded portion of the jungle, to die in solitude. Most animals have the same instinct, which induces them to seek the shelter of some spot remote from all disturbance; and should we find their remains, it will be near water, where the thirst of disease has been assuaged at the last moment. The ox tribe are subject to violent epidemics, and I have not only found the bodies of buffaloes in great numbers upon occasions during some malignant murrain, but they have been scattered throughout the country in all directions, causing a frightful stench, and probably extending the infection. A few years ago there was an epidemic among the bisons in the Reipore district of India; this spread into neighbouring districts over a large extent of country, and caused fearful ravages, but none of the deer tribe were attacked, the disease being confined specially to the genus Bos. There are interesting proofs of the specific poison of certain maladies which are limited in their action to a particular class of animal. We find the same in vegetable diseases, where a peculiar insect will attack a distinct family of plants, or where a special variety of fungoid growth exerts a similar baneful influence.

Wounded elephants have a marvellous power of recovery when in their wild state, although they have no gift of surgical knowledge, their simple system being confined to plastering their wounds with mud, or blowing dust upon the surface. Dust and mud comprise the entire pharmacopoeia of the elephant, and this is applied upon the most trivial as well as upon the most serious occasion. If an elephant has a very slight sore back, it will quickly point out the tender part by blowing dust with its trunk upon the spot which it cannot reach. Should the mahout have seriously punished the crown with the cruel driving-hook, the elephant applies dust at the earliest opportunity. I have seen them, when in a tank, plaster up a bullet-wound with mud taken from the bottom. This application is beneficial in protecting the wound from the attack of flies. The effect of these disgusting insects is quite shocking when an unfortunate animal becomes fly-blown, and is literally consumed by maggots. An elephant possesses a wonderful superiority over all other animals in the trunk, which can either reach the desired spot directly, or can blow dust upon it when required. All shepherds in England appreciate the difficulty when their sheep are attacked by flies, but they can be relieved by the human hand; a wild animal, on the contrary, has no alleviation, and it must eventually succumb to its misery. There is a peculiar fly in most tropical climates, but more especially in Ceylon, which lays live maggots, instead of eggs that require some time to hatch. These are the most dreadful pests, as the lively young maggots exhibit a horrible activity in commencing their work the instant they see the light; they burrow almost immediately into the flesh, and grow to a large size within twenty-four hours, occasioning the most loathsome sores. The best cure for any wound thus attacked, and swarming with live maggots, is a teaspoonful of calomel applied and rubbed into the deep sore.

I have seen the Arabs in the Soudan adopt a most torturing remedy when a camel has suffered from a fly-blown sore back. Upon one occasion I saw a camel kneeling upon the ground with a number of men around it, and I found that it was to undergo a surgical operation for a terrible wound upon its hump. This was a hole as large and deep as an ordinary breakfast-cup, which was alive with maggots. The operator had been preparing a quantity of glowing charcoal, which was at a red heat. This was contained in a piece of broken chatty, a portion of a water jar, and it was dexterously emptied into the diseased cavity on the camel's back.

The poor creature sprang to its feet, and screaming with agony, dashed at full gallop across the desert in a frantic state, with the fire scorching its flesh, and doubtless making it uncomfortable for the maggots. Fire is the Arabs' vade mecum; the actual cautery is deeply respected, and is supposed to be infallible. If internal inflammation should attack the patient, the surface is scored with a red-hot iron. Should guinea-worm be suspected, there is no other course to pursue than to burn the suffering limb in a series of spots with a red-hot iron ramrod. The worm will shortly make its appearance at one of these apertures after some slight inflammation and suppuration. This fearful complaint is termed Frendeet in the Soudan, and it is absorbed into the system generally by drinking foul water. At the commencement of the rainy season, when the ground has been parched by the long drought of summer, the surface-water drains into the hollows and forms muddy pools. The natives shun such water, as it is almost certain to contain the eggs of the guinea-worm. These in some mysterious manner are hatched within the body if swallowed in the act of drinking, and whether they develop in the stomach or in the intestines, it is difficult to determine, but the result is the same. The patient complains of rheumatic pains in one limb; this increases until the leg or arm swells to a frightful extent, accompanied by severe inflammation and great torment. The Arab practitioner declares that the worm is at work, and is seeking for a means of escape from the body. He accordingly burns half a dozen holes with a red-hot iron or ramrod. In a few days the head of the guinea-worm appears; it is immediately captured by a finely-split reed, and by degrees is wound like a cotton thread by turning the reed every day. This requires delicate manipulation, otherwise the worm might break, and a portion remain in the flesh, which would increase the inflammation. An average guinea-worm would be about three feet in length. Animals do not appear to suffer from this complaint, although they are subject to the attacks of great varieties of parasites. Elephants are frequently troubled with internal worms. I witnessed a curious instance of the escape of such insects from the stomach through a hole caused by a bullet, nevertheless the animal appeared to be in good condition.

It was a fine moonlight night on the borders of Abyssinia that I sat up to watch the native crops, which were a great attraction to the wild elephants, although there was no heavy jungle nearer than 20 miles. It was the custom of these animals to start after sunset, and to arrive at about ten o'clock in the vast dhurra fields of the Arabs, who, being without fire-arms, could only scare them by shouts and flaming torches. The elephants did not care much for this kind of disturbance, and they merely changed their position from one portion of the cultivated land to another more distant, and caused serious destruction to the crop (Sorghum vulgare), which was then nearly ripe. The land was rich, and the dhurra grew 10 or 12 feet high, with stems as thick as sugar-cane, while the large heads of corn contained several thousand grains the size of a split-pea. This was most tempting food for elephants, and they travelled nightly the distance named to graze upon the crops, and then retreated before sunrise to their distant jungles.

I do not enjoy night shooting, but there was no other way of assisting the natives, therefore I found myself watching, in the silent hours of night, in the middle of a perfect sea of cultivation, unbroken for many miles. There is generally a calm during the night, and there was so perfect a stillness that it was almost painful, the chirp of an insect sounding as loud as though it were a bird. At length there was a distant sound like wind, or the rush of a stream over a rocky bed. This might have been a sudden gust, but the sharp crackling of brittle dhurra stems distinctly warned us that elephants had invaded the field, and that they were already at their work of destruction.

As the dhurra is sown in parallel rows about 3 feet apart, and the ground was perfectly flat, there was no difficulty in approaching the direction whence the cracking of the dhurra could be distinctly heard. The elephants appeared to be feeding towards us with considerable rapidity, and in a few minutes I heard the sound of crunching within 50 yards of me. I immediately ran along the clear passage between the tall stems, and presently saw a black form close to me as it advanced in the next alley to my own. I do not think I was more than 4 or 5 yards from it when it suddenly turned its head to the right, and I immediately took a shot behind the ear. I had a white paper sight upon the muzzle of the large rifle (No. 10), which was plainly distinguished in the bright moonlight, and the elephant fell stone dead without the slightest struggle.

After some delay from the dispersion of my men who carried spare guns, I re-loaded, and followed in the direction which the herd had taken.

Although upon the "qui vive," they had not retreated far, as they were unaccustomed to guns, and they were determined to enjoy their supper after the long march of 20 miles to the attractive dhurra fields. I came up with them about three-quarters of a mile from the first shot; here there was the limit of cultivation, and all was wild prairie land; they had retreated by the way they had arrived, with the intention, no doubt, of returning again to the dhurra when the disturbing cause should have disappeared. I could see the herd distinctly as they stood in a compact body numbering some ten or twelve animals. The only chance was to run straight at them in order to get as near as possible before they should start, as I expected they would, in panic. Accordingly I ran forward, when, to my surprise, two elephants rushed towards me, and I was obliged to fire right and left. One fell to the ground for a moment, but recovered; the other made no sign, except by whirling round and joining the herd in full retreat.

That night I used a double-barrel muzzle-loader (No. 10), with conical bullet made of 12 parts lead, 1 part quicksilver, 7 drams of powder.

Some days later we heard native reports concerning an elephant that had been seen badly wounded, and very lame.

Forty-two days after this incident I had moved camp to a place called Geera, 22 miles distant. It was a wild uninhabited district at that time on the banks of the Settite river, with the most impervious jungle of hooked thorns, called by the Arabs "kittul." This tree does not grow higher than twenty-five feet, but it spreads to a very wide flat-topped head, the branches are thick, the wood immensely strong and hard, while the thorns resemble fish-hooks minus the barb. This impenetrable asylum was the loved resort of elephants, and it was from this particular station that they made their nocturnal raids upon the cultivated district more than 20 miles distant in a direct line.

We slept out that night upon the sandy bed of a small stream, which at that season of great heat had evaporated. Upon waking on the following morning we found the blankets wet through with the heavy dew, and the pillows soaking. Having arranged the camp, I left Lady Baker to give the necessary orders, while I took my rifles and a few good men for a reconnaissance of the neighbourhood.

The river ran through cliffs of rose-coloured limestone; this soon changed to white; and we proceeded down stream examining the sandy portions of the bed for tracks of game that might have passed during the preceding night. After about a mile we came upon tracks of elephants, which had apparently come down to drink at our side of the river, and had then returned, I felt sure, to the thorny asylum named Tuleet.

There was no other course to pursue but to follow on the tracks; this we did until we arrived at the formidable covert to which I have alluded. It was impossible to enter this except at certain places where wild animals had formed a narrow lane, and in one of these by-ways we presently found ourselves, sometimes creeping, sometimes walking, but generally adhering firmly every minute to some irrepressible branch of hooked thorns, which gave us a pressing invitation to "wait a bit." In a short time we found evident signs that the elephants were near at hand. The natives thrust their naked feet into the fresh dung to see if it was still warm. This was at length the case, and we advanced with extra care. The jungle became so thick that it was almost impossible to proceed. I wore a thick flaxen shirt which would not tear. This had short sleeves, as I was accustomed to bare arms from a few inches above the elbow. Not only my shirt, but the tough skin of my arms was every now and then hooked up fast by these dreadful thorns, and at last it appeared impossible to proceed. Just at that moment there was a sudden rush, a shrill trumpet, and the jungle crashed around us in magnificent style to those who enjoy such excitement, and a herd of elephants dashed through the dense thicket and consolidated themselves into a mighty block as they endeavoured to force down the tough thorny mass ahead of them. This was a grand opportunity to run in, but a phalanx of opposing rumps like the sterns of Dutch vessels in a crowd rendered it impossible to shoot, or to pass ahead of the perplexed animals. A female elephant suddenly wheeled round, and charged straight into us; fortunately I killed her with a forehead shot exactly below the boss or projection above the trunk. I now took a spare rifle, the half-pounder, and fired into the flank of the largest elephant in the herd, just behind the last rib, the shot striking obliquely, thus aimed to reach the lungs, as I could not see any of the fore portion of the body.

The dense compressed thorny mass of jungle offered such resistance that it was some time before it gave way before the united pressure of these immense animals. At length it yielded as the herd crashed through, but it then closed again upon us and made following impossible. However, we felt sure that the elephant I had hit with the half-pound explosive shell would die, and after creeping through upon the tracks with the greatest difficulty for about 150 yards, we found it lying dead upon its side.

The whole morning was occupied in cutting up the flesh and making a post-mortem examination. We found the inside partially destroyed by the explosive shell, which had shattered the lungs, but there was an old wound still open where a bullet had entered the chest, and missing the heart and lungs in an oblique course, it had passed through the stomach, then through the cavity of the body beneath the ribs and flank, and had penetrated the fleshy mass inside the thigh. In that great resisting cushion of strong muscles the bullet had expended its force, and found rest from its extraordinary course of penetration. After some trouble, I not only traced its exact route, but I actually discovered the projectile embedded in a foul mass of green pus, which would evidently have been gradually absorbed without causing serious damage to the animal. To my surprise, it was my own No. 10 two-groove conical bullet, composed of twelve parts lead and one of quicksilver, which I had fired when this elephant had advanced towards me at night, forty-two days ago, and 22 miles, as far as I could ascertain, from the spot where I had now killed it. The superior size of this animal to the remainder of the herd had upon both occasions attracted my special attention, hence the fact of selection, but I was surprised that any animal should have recovered from such a raking shot. The cavity of the body abounded with hairy worms about 2 inches in length. These had escaped from the stomach through the two apertures made by the bullet; and upon an examination of the contents, I found a great number of the same parasites crawling among the food, while others were attached to the mucous membrane of the paunch. This fact exhibits the recuperative power of an elephant in recovering from a severe internal injury.

The natives of Central Africa have a peculiar method of destroying them, by dropping a species of enormous dagger from the branch of a tree. The blade of this instrument is about 2 feet in length, very sharp on both edges, and about 3 inches in width at the base. It is secured in a handle about 18 inches long, the top of which is knobbed; upon this extremity a mass of well-kneaded tenacious clay mixed with chopped straw is fixed, weighing 10 or 12 lbs., or even more. When a large herd of elephants is discovered in a convenient locality, the hunt is thus arranged:—A number of men armed with these formidable drop-spears or daggers ascend all the largest and most shady trees throughout the neighbouring forest. In a great hunt there may be some hundred trees thus occupied. When all is arranged, the elephants are driven and forced into the forest, to which they naturally retreat as a place of refuge. It is their habit to congregate beneath large shady trees when thus disturbed, in complete ignorance of the fact that the assassins are already among the branches. When an elephant stands beneath a tree thus manned, the hunter drops his weighted spearhead so as to strike the back just behind the shoulder. The weight of the clay lump drives the sharp blade up to the hilt, as it descends from a height of 10 or 12 feet above the animal. Sometimes a considerable number may be beneath one tree, in which case several may be speared in a similar manner. This method of attack is specially fatal, as the elephants, in retreating through the forest, brush the weighted handle of the spear-blade against the opposing branches; these act as levers in cutting the inside of the animal by every movement of the weapon, and should this be well centred in the back there is no escape.

There is no animal that is more persistently pursued than the elephant, as it affords food in wholesale supply to the Africans, who consume the flesh, while the hide is valuable for shields; the fat when boiled down is highly esteemed by the natives, and the ivory is of extreme value. No portion of the animal is wasted in Africa, although in Ceylon the elephant is considered worthless, and is allowed to rot uselessly upon the ground where it fell to die.

The professional hunters that are employed by European traders shoot the elephant with enormous guns, or rifles, which are generally rested upon a forked stick driven into the ground. In this manner they approach to about 50 yards' distance, and fire, if possible simultaneously, two shots behind the shoulder. If these shots are well placed, the elephant, if female, will fall at once, but if a large male, it will generally run for perhaps 100 or more yards until it is forced to halt, when it quickly falls, and dies from suffocation, if the lungs are pierced.

The grandest of all hunters are the Hamran Arabs, upon the Settite river, on the borders of Abyssinia, who have no other weapon but the heavy two-edged sword. I gave an intimate account of these wonderful Nimrods many years ago in the *Nile Tributaries of Abyssinia*, but it is impossible to treat upon the elephant without some reference to these extraordinary people.

Since I visited that country in 1861, the published account of those travels attracted several parties of the best class of ubiquitous Englishmen, and I regret to hear that all those mighty hunters who accompanied me have since been killed in the desperate hand-to-hand encounters with wild elephants. Their life is a constant warfare with savage beasts, therefore it may be expected that the termination is a death upon their field of battle, invariably sword in hand.

James Bruce, the renowned African traveller of the last century, was the first to describe the Agagheers of Abyssinia, and nothing could be more graphic than his description both of the people and the countries they inhabit, through which I have followed in Bruce's almost forgotten footsteps, with the advantage of possessing his interesting book as my guide wheresoever I went in 1861. Since that journey, the deplorable interference of England in Egypt which resulted in the abandonment of the Soudan and the sacrifice of

General Gordon at Khartoum has completely severed the link of communication that we had happily established established, which had laid the foundations for future civilisation. The splendid sword-hunters of the Hamran Arabs, who were our friends in former days, have been converted into enemies by the meddling of the British Government with affairs which they could not understand. It is painful to look back to the past, when Lady Baker and myself, absolutely devoid of all escort, passed more than twelve months in exploring the wildest portions of the Soudan, attended only by one Egyptian servant, assisted by some Arab boys which we picked up in the desert among the Arab tribes. In those days the name of England was respected, although not fairly understood. There was a vague impression in the Arab mind that it was the largest country upon earth; that its Government was the emblem of perfection; that the military power of the country was overwhelming (having conquered India); and that the English people always spoke the truth, and never forsook their friends in the moment of distress. There was also an idea that England was the only European Power which regarded the Mussulmans with a friendly eye, and that, were it not for British protection, the Russians would eat the Sultan and overthrow the mosques, to trample upon the Mahommedan power in Constantinople. England was therefore regarded as the friend and the ally of the Mahommedans; it was known that we had together fought against the Russians, and it was believed that we were always ready to fight in the same cause when called upon by the Sultan. All British merchandise was looked upon as the ne plus ultra of purity and integrity; there could be no doubt of the quality of goods, provided that they were of English manufacture.

An Englishman cannot show his face among those people at the present day. The myth has been exploded. The golden image has been scratched, and the potter's clay beneath has been revealed. This is a terrible result of clumsy management. We have failed in every way. Broken faith has dissipated our character for sincerity, and our military operations have failed to attain their object, resulting in retreat upon every side, to be followed up even to the seashores of the Red Sea by an enemy that is within range of our gun-vessels at Souakim. This is a distressing change to those who have received much kindness and passed most agreeable days among the

Arab tribes of the Soudan deserts, and I look back with intense regret to the errors we have committed, by which the entire confidence has been destroyed which formerly was associated with the English name. The countries which we opened by many years of hard work and patient toil throughout the Soudan, even through the extreme course of the White Nile to its birthplace in the equatorial regions, have been abandoned by the despotic order of the British Government, influenced by panic instead of policy; telegraphic lines which had been established in the hitherto barbarous countries of Kordofan, Darfur, the Blue Nile territories of Senaar, and throughout the wildest deserts of Nubia to Khartoum have all been abandoned to the rebels, who under proper management should have become England's friends.

This has been our civilising influence (?), by which we have broken down the work of half a century, and produced the most complete anarchy where five-and-twenty years ago a lady could travel in security. England entered Egypt in arms to *re-establish the authority of the Khedive!* We have dislocated his Empire, and forsaken the Soudan.

CHAPTER IV

The experience of modern practice has hardly decided the vexed question "whether the African species is more difficult to train than the gentle elephant of Asia." In a wild state there can be no doubt that the African is altogether a different animal both in appearance and in habits; it is vastly superior in size, and although of enormous bulk, it is more active and possesses greater speed than the Asiatic variety. Not only is the marked difference in shape a distinguishing peculiarity,—the hollow back, the receding front, the great size of the ears,—but the skin is rougher, and more decided in the bark-like appearance of its texture.

The period of gestation is considered to be the same as the Asiatic elephant, about twenty-two months, but this must be merely conjecture, as there has hitherto been no actual proof. My own experience induces me to believe that the African elephant is more savage, and although it may be tamed and rendered docile, it is not so dependable as the Asiatic. Only last year I saw an African female in a menagerie who had killed her keeper, and was known to be most treacherous. Her attendant informed me that she was particularly fond of change, and would welcome a new keeper with evident signs of satisfaction, but after three or four days she would tire of his society and would assuredly attempt to injure him, either by backing and squeezing him against the wall, or by kicking should he be within reach of her hind legs.

Few persons are aware of the extreme quickness with which an elephant can kick, and the great height that can be reached by this mischievous use of the hind foot. I have frequently seen an elephant kick as sharp as a small pony, and the effect of a blow from so ponderous a mass propelled with extreme velocity may be imagined. This is a peculiar action, as the elephant is devoid of hocks, and it uses the knees of the hind legs in a similar manner to those of a human being, therefore a backward kick would seem unnatural; but the elephant can kick both backwards and forwards with equal dexterity, and this constitutes a special means of defence against an enemy, which seldom escapes when exposed to such a game between the fore and hind feet of the infuriated animal.

Although it is generally believed that an elephant moves the legs upon each side simultaneously, like the camel, it does not actually touch the ground with each foot upon the same side at exactly the same moment, but the fore foot touches the surface first, rapidly followed by the hind, and in both cases the heel is the first portion of the foot that reaches its destination. The effect may be seen in the feet of an elephant after some months' continual marching upon hard ground: the heels are worn thin and are quite polished, as though they had been worn down by the friction of sand-paper,-in fact, they are in the same condition as the heels of an old boot.

The Indian native princes do not admire the African elephant, as it combines many points which are objectionable to their peculiar ideas of elephantine proportions. According to their views, the hollow back of an African elephant would amount to a deformity. The first time that I ever saw a large male of that variety I was of the same opinion. I was hunting with the Hamran Arabs in a wild and uninhabited portion of Abyssinia, along the banks of the Settite river, which is the main stream of the Atbara, the chief affluent of the Nile.

As before stated, I have already published an account of these wonderful hunters in the Nile Tributaries of Abyssinia, and it is sufficient to describe them as the most fearless and active followers of the chase, armed with no other weapon than the long, straight, two-edged Arab sword, with which they attack all animals, from the elephant and rhinoceros to the lion and buffalo. The sword is sharpened to the finest degree, and the blade is protected for about six inches above the cross-hilt with thick string, bound tightly round so as to afford a grip for the right hand, while the left grips the hilt in the usual manner. This converts the ordinary blade into a two-handed sword, a blow from which will sever a naked man into two halves if delivered at the waist. It may be imagined that a quick cut from such a formidable weapon will at once divide the hamstring of any animal. The usual method of attacking the elephant is as follows:-Three, or at the most four mounted hunters sally forth in quest of game. When the fresh tracks of elephants are discovered they are steadily followed up until the herd, or perhaps the single animal, is found. If a large male with valuable tusks, it is singled out and separated from the herd. The leading hunter follows the retreating elephant, accompanied by his companions in single file. After a close hunt, keeping within 10 yards of the game, a sudden halt becomes necessary, as the elephant turns quickly round and faces its pursuers.

The greatest coolness is required, as the animal, now thoroughly roused, is prepared to charge. The hunters separate to right and left, leaving the leader to face the elephant. After a few moments, during which the hunter insults the animal by shouting uncomplimentary remarks concerning the antecedents of its mother, and various personal allusions to imaginary members of the family, the elephant commences to back a half-dozen paces as a preliminary to a desperate onset. This is the well-known sign of the coming charge. A sharp shrill trumpet! and, with its enormous ears thrown forward, the great bull elephant rushes towards the apparently doomed horse. As quick as lightning the horse is turned, and a race commences along a course terribly in favour of the elephant, where deep ruts, thick tangled bush, and the branches of opposing trees obstruct both horse and rider. Everything now depends upon the sure-footedness of the horse and the cool dexterity of the rider. For the first 100 yards an elephant will follow at 20 miles an hour, which keeps the horse flying at top speed before it. The rider, even in this moment of great danger, looks behind him, and adapts his horse's pace so narrowly to that of his pursuer that the elephant's attention is wholly absorbed by the hope of overtaking the unhappy victim.

In the meantime, two hunters follow the elephant at full gallop; one seizes his companion's reins and secures the horse, while the rider springs to the ground with the same agility as a trained circus-rider, and with one dexterous blow of his flashing sword he divides the back sinew of the elephant's hind leg about 16 inches above the heel. The sword cuts to the bone. The elephant that was thundering forward at a headlong speed suddenly halts; the foot dislocates when the great weight of the animal presses upon it deprived of the supporting sinew. That one cut of the sharp blade, disables an animal which appeared invincible.

As the elephant moves both legs upon the same side simultaneously, the disabling of one leg entirely cripples all progress, and the creature becomes absolutely helpless. The hunter, having delivered his fatal stroke, springs nimbly upon one side to watch the effect, and then without difficulty he slashes the back sinew of the remaining leg, with the result that the animal bleeds to death. This is a cruel method, but it requires the utmost dexterity and daring on the part of the hunters, most of whom eventually fall victims to their gallantry.

I was accompanied by these splendid sword-hunters of the Hamran Arabs in 1861 during my exploration of the Nile tributaries of Abyssinia; and upon the first occasion that I was introduced to an African male elephant, the animal was standing at the point of a long sandbank which had during high water formed the bed of the river, where a sudden bend had hollowed out the inner side of the curve and thrown up a vast mass of sand upon the opposite shore. This bank was a succession of terraces, each about 4 feet high, formed at intervals during the changes in the level of the retreating stream. The elephant was standing partly in the water drinking, and quite 100 yards from the forest upon the bank. The huge dark mass upon the glaring surface of white sand stood out in bold relief and exhibited to perfection the form and proportions of the animal; but it was so unlike the Indian elephant of my long experience that I imagined some accident must have caused a deformity of the back, which was deeply hollowed, instead of being convex like the Asiatic species. I whispered this to my hunters, who did not seem to understand the remark; and they immediately dismounted, exclaiming that the loose sand was too deep for their horses, and they preferred to be on foot.

It was difficult to approach this elephant, as there was no cover whatever upon the large area of barren sand; the only method was to keep close to the level of the water below the terraces, as the head of the animal was partially turned away from us whilst drinking. I had a very ponderous single rifle weighing 22 lbs., which carried a conical shell of half a pound, with a charge of 16 drams of powder. The sand was so deep that any active movement would have been impossible with the load of so heavy a weapon; I therefore determined to take a shoulder shot should I be able to arrive unperceived within 50 yards. Stooping as low as possible, and occasionally lying down as the ever-swinging head moved towards us, we at length arrived at the spot which I had determined upon for the fatal shot. Just at that moment the elephant perceived us, but before he had made up his mind, I fired behind the shoulder, and as the smoke cleared, I distinctly saw the bullet-hole, with blood flowing from the wound. I think the elephant would have charged, but without a moment's hesitation my gallant Hamrans rushed towards him sword in hand in the hope of slashing his hamstring before he could reach the forest. This unexpected and determined onset decided the elephant to retreat, which he accomplished at such a pace, owing to the large surface of his feet upon the loose sand, that the active hunters were completely distanced, although they exerted themselves to the utmost in their attempts to overtake him.

The wound through the shoulder was fatal, and the elephant fell dead in thick thorny jungle, to which it had hurried as a secure retreat. This was a very large animal, but as I did not actually measure it, any guess at the real height would be misleading. As before noted, the measurement of the African elephant Jumbo, when sold by the Zoological Society of London, was 11 feet in height of shoulder, and 6 tons 10 cwts. nett when weighed before shipment at the docks. That animal might be accepted as a fair specimen, although it would be by no means unusual to see wild elephants which greatly exceed this size.

The peculiar shape of head renders a front shot almost impossible, and the danger of hunting the African elephant is greatly enhanced by this formation of the skull, which protects the brain and offers no defined point for aim.

I have never succeeded in killing a male African elephant by the forehead shot, although it is certainly fatal to the Asiatic variety if placed rather low, in the exact centre of the boss or projection above the trunk. Should an African elephant charge, there is no hope of killing the animal by a direct shot, and the only chance of safety for the hunter is the possession of good nerves and a powerful double-barrelled rifle, No. 8 or No. 4, with 14 drams of powder and a well-hardened bullet. The right-hand barrel will generally stop a charging elephant if the bullet is well placed very low, almost in the base of the trunk. Should this shot succeed in turning the animal, the left-hand barrel would be ready for a shot in the exact centre of the shoulder; after which, time must be allowed for the elephant to fall from internal haemorrhage.

There is no more fatal policy in hunting dangerous game than a contempt of the animal, exhibited by a selection of weapons of inferior calibre. Gunmakers in London of no practical experience, but who can only trust to the descriptions of those who have travelled in wild countries, cannot possibly be trusted as advisers. Common sense should be the guide, and surely it requires no extraordinary intelligence to understand that a big animal requires a big bullet, and that a big bullet requires a corresponding charge of powder, which necessitates a heavy rifle. If the hunter is not a Hercules, he cannot wield his club; but do not permit him to imagine that he can deliver the same knock-down blow with a lighter weapon, simply because he cannot use the heavier.

We lost only last year one of the most daring and excellent men, who was an excellent representative of the type which is embraced in the proud word "Englishman"—Mr. Ingram—who was killed by a wild female elephant in Somali-land, simply because he attacked the animal with a '450 rifle. Although he was mounted, the horse would not face some prickly aloes which surrounded it, and the elephant, badly but not really seriously wounded, was maddened by the attack, and, charging home, swept the unfortunate rider from his saddle and spitted him with her tusks.

This year (1889) we have to lament the death of another fine specimen of our countrymen, the Hon. Guy Dawnay, who has been killed by a wild buffalo in East Africa. The exact particulars will never be ascertained, but it appears that he was following through thick jungle a wounded buffalo, which suddenly turned and was not stopped by the rifle.

I cannot conceive anything more dangerous than the attack of such animals with an inferior weapon. Nothing is more common than the accounts of partially experienced beginners, who declare that the '450 bore is big enough for anything, because they have happened to kill a buffalo or rhinoceros by a shoulder shot with such an inferior rifle. If the animal had been facing them, it would have produced no effect whatever, except to intensify the charge by maddening the already infuriated animal.

This is the real danger in the possession of what is called a " handy small-bore," when in wild countries abounding in dangerous game. You are almost certain to select for your daily companion the lightest and handiest rifle, in the same manner that you may use some favourite walking-stick which you instinctively select from the stand that is filled with a variety.

All hunters of dangerous animals should accustom themselves to the use of large rifles, and never handle anything smaller than a '577, weighing 12 lbs., with a solid 650 grain hard bullet, and at the least 6 drams of powder. I impress this upon all who challenge the dangers of the chase in tropical climates. No person of average strength will feel the weight of a 12 lb. rifle when accustomed to its use. Although this is too small as a rule for heavy game, it is a powerful weapon when the bullet is hardened by a tough mixture of antimony or quicksilver. A shoulder shot from such a rifle will kill any animal less than an elephant, and the front shot, or temple, or behind the ear, will kill any Asiatic elephant.

I would not recommend so small a bore for heavy thick-skinned game, but the '577 rifle is a good protector, and you need not fear any animal in your rambles through the forest when thus armed, whereas the '450 and even the '500 would be of little use against a charging buffalo.

At the same time it must be distinctly understood that so light a projectile as 650 grains will not break the bone of an elephant's leg, neither will it penetrate the skull of a rhinoceros unless just behind the ear. This is sufficient to establish the inferiority of small-bores.

I have seen in a life's experience the extraordinary vagaries of rifle bullets, and for close ranges of 20 yards there is nothing, in my opinion, superior to the old spherical hardened bullet with a heavy charge of powder. The friction is minimised, the velocity is accordingly increased, and the hard round bullet neither deflects nor alters its form, but it cuts through intervening branches and goes direct to its aim, breaking bones and keeping a straight course through the animal. This means death.

At the same time it must be remembered that a '577 rifle may be enabled to perform wonders by adapting the material of the bullet to the purpose specially desired. No soft-skinned animal should be shot with a hardened bullet, and no hard-skinned animal should be shot with a soft bullet.

You naturally wish to kill your animal neatly — to double it up upon the spot. This you will seldom or never accomplish with a very hard bullet and a heavy charge of powder, as the high velocity will drive the hard projectile so immediately through the animal that it receives no striking energy, and is accordingly unaware of a fatal wound that it may have received, simply because it has not sustained a shock upon the impact of a bullet which has passed completely through its body.

To kill a thin-skinned animal neatly, such as a tiger, lion, large deer, etc. etc., the bullet should be pure lead, unmixed with any other metal. This will flatten to a certain degree immediately upon impact, and it will continue to expand as it meets with resistance in passing through the tough muscles of a large animal, until it assumes the shape of a fully developed mushroom, which, after an immense amount of damage in its transit, owing to its large diameter, will remain fixed beneath the skin upon the side opposite to its place of entry. This bestows the entire striking energy of the projectile, and the animal succumbs to the tremendous shock, which it would not have felt had the bullet passed through, carrying on its striking energy until stopped by some other object beyond.

I must repeat that although gunmakers object to the use of pure lead for rifle bullets, upon the plea that lead will form a coating upon the inner surface of the barrel, and that more accurate results will be obtained in target practice by the use of hardened metal, the argument does not apply to sporting practice. You seldom fire more than half a dozen shots from each barrel during the day, and the rifle is well cleaned each evening upon your return to camp. The accuracy with a pure leaden bullet is quite sufficient for the comparatively short ranges necessitated by game-shooting. The arguments of leading the barrel, etc., cannot be supported, and the result is decidedly in favour of pure lead for all soft-skinned animals.

The elephant requires not only a special rifle, but the strongest ammunition that can be used without injury to the shooter by recoil. It is impossible to advocate any particular size of rifle, as it must depend upon the strength of the possessor. As a rule I do not approve of shells, as they are comparatively useless if of medium calibre, and can be only effective when sufficiently large to contain a destructive bursting charge. I have tried several varieties of shells with unsatisfactory results, excepting the half-pounder, which contained a burst bursting charge of 8 drams of the finest grained powder.

This pattern was my own invention, as I found by experience that the general defect of shells was the too immediate explosion upon impact. This would cause extensive damage to the surface, but would fail in penetration.

Picrate of potash was at one time supposed to combine an enormous explosive power with perfect safety in carriage, as the detonating shells were proof against the blow of a hammer, and would only explode upon impact through the extreme velocity of their discharge from a rifle-barrel. These were useless against an elephant, as they had no power of penetration, and the shell destroyed itself by bursting upon the hard skin. I tried these shells against trees, but although the bark would be shattered over an extensive area, upon every occasion the projectile failed to penetrate the wood, as it had ceased to exist upon explosion on the surface.

My half-pound shell was exceedingly simple. A cast-iron bottle, similar in shape to a German seltzer-water, formed the core, around which the lead was cast. The neck of the iron bottle projected through the pointed cone of the projectile, and formed a nipple to receive the percussion-cap. In external appearance the shell was lead, the iron bottle being concealed within. Half an ounce of the finest grained powder was inserted through the nipple by means of a small funnel; this formed the bursting charge. The cap was only adjusted previous to loading, as a necessary precaution. This half-pound shell was propelled by a charge of 16 drams of coarse-grained powder.

I never fired this rifle without killing the animal, but the weapon could not be claimed as a pleasant companion, the recoil being terrific. The arrangement of the cap upon a broad-mouthed nipple prevented the instantaneous explosion that would have taken place with a picrate of potash shell. A fraction of a second was required to explode the cap upon impact, and for the cap to ignite the bursting charge; this allowed sufficient time for the shell to penetrate to the centre of an elephant before the complete ignition had taken place. The destruction occasioned by the half-ounce of powder confined within the body of an elephant may be imagined.

I tried this shell at the forehead of a hippopotamus, which was an admirable test of penetration before bursting. It went through the brain, knocked out the back of the skull, and exploded within the neck, completely destroying the vertebrae of the spine, which were reduced to pulp, and perforating a tunnel blackened with gunpowder several feet in length, along which I could pass my arm to the shoulder. The terminus of the tunnel contained small fragments of lead and iron, pieces of which were found throughout the course of the explosion.

The improvements in modern rifles will, within the next half-century, be utterly destructive to the African elephant, which is unprotected by laws in the absence of all government. For many ages these animals have contended with savage man in unremitting warfare, but the lance and arrow have been powerless to exterminate, and the natural sagacity of the elephant has been sufficient to preserve it from wholesale slaughter among pitfalls and other snares. The heavy breechloading rifle in the hands of experienced hunters is a weapon which nothing can withstand, and the elephants will be driven far away into the wilderness of an interior where they will be secure from the improved fire-arms of our modern civilisation.

It is much to be regretted that no system has been organised in Africa for capturing and training the wild elephants, instead of harrying them to destruction. In a country where beasts of burden are unknown, as in equatorial Africa, it appears incredible that the power and the intelligence of the elephant have been completely ignored. The ancient coins of Carthage exhibit the African elephant, which in those remote days was utilised by the Carthaginians; but a native of Africa, if of the Negro type, will never tame an animal, he only destroys.

When we consider the peculiar power that an elephant possesses for swimming long distances, and for supporting long marches under an enormous weight, we are tempted to condemn the apathy even of European settlers in Africa, who have hitherto ignored the capabilities of this useful creature. The chief difficulty of African commerce is the lack of transport. The elephant is admirably adapted by his natural habits for travelling through a wild country devoid of roads. He can wade through unbridged streams, or swim the deepest rivers (without a load), and he is equally at home either on land or water. His carrying power for continued service would be from 12 to 14 cwts.; thus a single elephant would convey about 1300 lbs. of ivory in addition to the weight of the pad. The value of one load would be about 5oo pounds. At the present moment such an amount of ivory would employ twenty-six carriers; but as these are generally slaves which can be sold at the termination of the journey, they might be more profitable than the legitimate transport by an elephant.

Although the male elephant will carry a far greater load than the female, through its superior size and strength, it would be dangerous to manage upon a long journey should it take place during the period, of "must." I have heard the suggestion that an elephant should be castrated, as the operation would affect the temper of the animal and relieve it from the irritation of the "must" period; but such an operation would be impossible, as the elephant is peculiarly formed, and, unlike other animals, it has neither scrotum nor testicles externally. These are situated within the body, and could not be reached by surgery.

It is well known that the entire males of many domestic animals are naturally savage. The horse, bull, boar, and the park-fed stag are all uncertain in their tempers and may be pronounced unsafe; but the male elephant, although dangerous to a stranger and treacherous to his attendants, combines an extraordinary degree of cowardice with his natural ferocity. A few months ago I witnessed a curious example of this combination in the elephant's character. A magnificent specimen had been lent to me by the Commissariat Department at Jubbulpur; this was a high caste bull elephant named Bisgaum that was well known as bad-tempered, but was supposed to be courageous. He had somewhat tarnished his reputation during the last season by turning tail upon a tiger that rushed out of dense bush and killed a coolie within a few yards of his trunk; but this momentary panic was excused, and the blame was thrown upon the mahout. The man was dismissed, and a first-rate Punjaubi driver was appointed in his stead. This man assured me that the elephant was dependable; I accordingly accepted him, and he was ordered to carry the howdah throughout the expedition.

In a very short experience we discovered the necessity of giving Bisgaum a wide berth, as he would fling out his trunk with extreme quickness to strike a person within his reach, and he would kick out sharply with his hind leg whenever a native ventured to approach his rear. He took a fancy to me, as I fed him daily with sugar-canes, jaggery, and native chupatties (cakes), which quickly established an understanding between us; but I always took the precaution of standing by his side instead of in his front, and of resting my left hand upon his tusk while I fed him with the right. Every morning at daylight he was brought to the tent with Demoiselle (the female elephant), and they both received from my own hands the choice bits which gained their confidence.

My suspicions were first aroused by his peculiar behaviour upon an occasion when we had killed two tigers; these were young animals, and although large, there was no difficulty in arranging them upon the pad, upon which they were secured by ropes, when the elephant kneeling down was carefully loaded. Hardly had Bisgaum risen to his feet, when, conscious of the character of the animals upon his back, and, I suppose, not quite certain that life was actually extinct, he trumpeted a shrill scream, and shook his immense carcase like a wet dog that has just landed from the water. This effect was so violent that one tiger was thrown some yards to the right, while the other fell to the ground on the left, and without a moment's warning, the elephant charged the lifeless body, sent it flying by a kick with his fore foot, and immediately proceeded to dance a war-dance, kicking with his hind legs to so great a height that he could have reached a tall man's hat. A vigorous application of the driving-hook by the mahout, who was a powerful man, at length changed the scene, and the elephant at once desisted from his attack upon the dead tiger, and rushed madly upon one side, where he stood nervously looking at the enemy as though he expected it would show signs of life.

This did not look promising for an encounter with a live tiger, as it would have been absolutely impossible to shoot from that elephant's back.

A short time after this occurrence, when upon my usual reconnaissance through the jungles in the neighbourhood of the camp, I came upon the fresh tracks of a large tiger close to the banks of the Bearmi river, and I gave the necessary instructions that a buffalo should be tied up as a bait that same evening.

Early on the following morning the news was brought by the shikaris that the buffalo had been killed, and dragged into a neighbouring ravine. As the river was close by, there could be no doubt that the tiger would have drunk water after feasting on the carcase, and would be lying asleep somewhere in the immediate neighbourhood.

The mucharns (platforms in trees) had already been prepared in positions where the tiger was expected to pass when driven, as he would make for the forest-covered hills which rose within half a mile of the river.

The spot was within twenty minutes of the camp; the elephants were both ready, with simple pads, as the howdah was ill-adapted for a forest; and we quickly started.

Three mucharns had been prepared; these were about 100 yards apart in a direct line which guarded a narrow glade between the jungle upon the river's bank and the main body of the forest at the foot of a range of red-sandstone hills; these were covered to the summit with trees already leafless from the drought.

The mucharn which fell to my share was that upon the right flank when facing the beat; this was in the open glade opposite a projecting corner of the jungle. On the left, about 70 yards distant, was a narrow strip of bush connected with the jungle, about 4 yards wide, which terminated in a copse about 30 yards in diameter; beyond this was open glade for about 40 yards width until it bounded the main forest at the foot of the hill-range.

We took our places, and I was assured by the shikaris that the tiger would probably break covert exactly in my front.

It is most uncomfortable for a European to remain squatted in a mucharn for any length of time; the limbs become stiffened, and the cramped position renders good shooting anything but certain. I have a simple wooden turnstool, which enables me to shoot in any required direction; this is most comfortable.

I had adjusted my stool upon a thick mat to prevent it from slipping, and having settled myself firmly, I began to examine the position to form an opinion concerning the most likely spot for the tiger to emerge from the jungle.

The beat had commenced, and the shouts and yells, of a long line of 150 men were gradually becoming more distinct. Several peacocks ran across the open glade: these birds are always the forerunners of other animals, as they are the first to retreat.

Presently I heard a rustle in the jungle, and I observed the legs of a sambur deer, which, having neared the edge, now halted to listen to the beaters before venturing to break from the dense covert. The beaters drew nearer, and a large doe sambur, instead of rushing quickly forward, walked slowly into the open, and stood within 10 yards of me upon the glade. She waited there for several minutes, and then, as if some suspicion had suddenly crossed her mind, gave two or three convulsive bounds and dashed back to the same covert from which she had approached.

It struck me that the sambur had got the wind of an enemy, otherwise she would not have rushed back in such sudden haste; she could not have scented me, as I was 10 or 12 feet above the ground, and the breeze was aslant . . . . Then, if a tiger were in the jungle, why should she dash back into the same covert ?

I was reflecting upon these subjects, and looking out sharp towards my left and front, when I gently turned upon my stool to the right; there was the tiger himself! who had already broken from the jungle about 75 yards from my position. He was slowly jogging along as though just disturbed (possibly by the sambur), keeping close to the narrow belt of bushes already described. There was a foot-path from the open glade which pierced the belt; I therefore waited until he should cross this favourable spot. I fired with the '577 rifle just as he was passing across the dusty track. I saw the dust fly from the ground upon the other side as the hardened bullet passed like lightning through his flank, but I felt that I was a little too far behind his shoulder, as his response to the shot was a bound at full gallop forwards into the small clump of jungle that projected into the grassy open. My turnstool was handy, and I quickly turned to the right, waiting with the left-hand barrel ready for his re-appearance upon the grass-land in the interval between the main jungle and the narrow patch. There was no time to lose, for the tiger appeared in a few seconds, dashing out of the jungle, and flying over the open at tremendous speed. This was about 110 yards distant; aiming about 18 inches in his front, I fired. A short but spasmodic roar and a sudden convulsive twist of his body showed plainly that he was well hit, but with unabated speed he gained the main forest which was not more than 40 yards distant. If that had been a soft leaden bullet he would have rolled over to the shot, but I had seen the dust start from the ground when I fired, and I knew that the hard bullet had passed through without delivering the shock required.

The beaters and shikaris now arrived, and having explained the incident, we examined the ground for tracks, and quickly found the claw-marks which were deeply indented in the parched surface of fine sward. We followed these tracks cautiously into the jungle. Our party consisted of Colonel Lugard, the Hon. D. Leigh, myself, and two experienced shikaris. Tiger-shooting is always an engrossing sport, but the lively excitement is increased when you follow a wounded tiger upon foot. We now slowly advanced upon the track, which faintly showed the sharp claws where the tiger had alighted in every bound. The jungle was fairly open, as the surface was stony, and the trees for want of moisture in a rocky soil had lost their leaves; we could thus see a considerable distance upon all sides. In this manner we advanced about 100 yards without finding a trace of blood, and I could see that some of my people doubted the fact of the tiger being wounded. I felt certain that he was mortally hit, and I explained to my men that the hard bullet would make so clean a hole through his body that he would not bleed externally until his inside should be nearly full of blood. Suddenly a man cried "koon" (blood), and he held up a large dried leaf of the teak-tree upon which was a considerable red splash: almost immediately after this we not only came upon a continuous line of blood, but we halted at a place where the animal had lain down; this was a pool of blood, proving that the tiger would not be far distant.

I now sent for the elephants, as I would not permit the shikaris to advance farther upon foot. The big tusker Bisgaum arrived, and giving my Paradox gun to my trustworthy shikari Kerim Bux, he mounted the pad of that excitable beast to carry out my orders, "to follow the blood until he should find the tiger, after which he was to return to us." We were now on the top of a small hill within an extensive forest range, and directly in front the ground suddenly dipped, forming a V-shaped dell, which in the wet season was the bed of a considerable torrent. It struck me that if the tiger were still alive he would steal away along the bottom of the rocky watercourse; therefore, before the elephant should advance, and perhaps disturb him, we should take up a position on the right to protect the nullah or torrent-bed; this plan was accordingly carried out.

We had not been long in our respective positions when a shot from the direction taken by the elephant, followed instantly by a short roar, proved that the tiger had been discovered, and that he was still alive. My female elephant Demoiselle, upon hearing the sound, trembled beneath me with intense excitement, while the other female would have bolted had she not been sharply reminded by the heavy driving-hook. Several shots were now fired in succession, and after vainly endeavouring to discover the whereabouts of the tiger, I sent Demoiselle to obtain the news while we kept guard over the ravine. No tiger having appeared, I stationed natives in trees to watch the nullah while we ascended the hill on foot, directing our course through the forest to the place from whence the shots had been fired. We had hardly advanced 80 yards before we found both the elephants on the top of the steep shoulder of the hill, where several of our men were upon the boughs of surrounding trees. Bisgaum was in a state of wild excitement, and Kerim Bux explained that it was impossible to shoot from his back, as he could not be kept quiet. Where was the tiger? That was the question. "Close to us, Sahib!" was the reply; but on foot we could see nothing, owing to high withered grass and bush. I clambered upon the back of the refractory Bisgaum, momentarily expecting him to bolt away like a locomotive engine, and from that elevated position I was supposed to see the tiger, which was lying in the bottom of the ravine about 100 yards distant. There were so many small bushes and tufts of yellow grass that I could not distinguish the form for some minutes; at length my eyes caught the object. I had been looking for orange and black stripes, therefore I had not noticed black and white, the belly being uppermost, as the animal was lying upon its back, evidently dying.

The side of the rocky hill was so steep and slippery that the elephants could not descend; I therefore changed my steed and mounted Demoiselle, from the back of which I fired several shots at the tiger until life appeared to be extinct. The ground was so unfavourable that I would not permit any native to approach near enough to prove that the animal was quite dead. I therefore instructed Bisgaum's mahout to make a detour to the right until he could descend with his elephant into the flat bottom of the watercourse, he was then to advance cautiously until near enough to see whether the tiger breathed. At the same time I rode Demoiselle carefully as near as we could safely descend among the rocks to a distance of about 40 yards; it was so steep that the elephant was impossible to turn. From this point of vantage I soon perceived Bisgaum's bulky form advancing up the dry torrent-bed. The rocks were a perfectly flat red sandstone, which in many places resembled artificial pavement; this was throughout the district a peculiar geological feature, the surface of the stone being covered with ripple-marks, and upon this easy path Bisgaum now approached the body of the tiger, which lay apparently dead exactly in his front.

Suddenly the elephant halted when about 15 yards from the object, which had never moved. I have seen wild savages frenzied by the exciting war-dance, but I never witnessed such an instance of hysterical fury as that exhibited by Bisgaum. It is impossible to describe the elephantine antics of this frantic animal; he kicked right and left with his hind legs alternately, with the rapidity of a horse; trumpeting and screaming, he threw his trunk in the air, twisting it about, and shaking his immense head, until, having lashed himself into sufficient rage, he made a desperate charge at the supposed defunct enemy, with the intention of treating the body in a similar manner to that a few days previous. But the tiger was not quite dead and although he could not move to get away, he seized with teeth and claws the hind leg of the maddened elephant, who had clumsily overrun him in the high excitement, instead of kicking the body with a fore foot as he advanced.

The scene was now most interesting. We were close spectators looking down upon the exhibition as though upon an arena. I never saw such fury in an elephant; the air was full of stones and dust, as he kicked with such force that the tiger for the moment was lost to view in the tremendous struggle, and being kicked away from his hold, with one of his long fangs broken short off to the gum, he lay helpless before his huge antagonist, who, turning quickly round, drove his long tusks between the tiger's shoulders, and crushed the last spark of life from his tenacious adversary.

This was a grand scene, and I began to think there was some real pluck in Bisgaum after all, although there was a total want of discipline; but just as I felt inclined to applaud, the victorious elephant was seized with a sudden panic, and turning tail, he rushed along the bottom of the watercourse at the rate of 20 miles an hour, and disappeared in the thorny jungle below at a desperate pace that threatened immediate destruction to his staunch mahout. Leaving my men to arrange a litter with poles and cross-bars to carry the tiger home, I followed the course of Bisgaum upon Demoiselle, expecting every minute to see the body of his mahout stretched upon the ground.

At length, after about half a mile passed in anxiety, we discovered Bisgaum and his mahout both safe upon an open plain; the latter torn and bleeding from countless scratches while rushing through the thorny jungle.

On the following day the elephant's leg was much swollen, although the wounds appeared to be very slight. It is probable that a portion of the broken tooth remained in the flesh, as the leg festered, and became so bad that the elephant could not travel for nearly a fortnight afterwards. The mahouts are very obstinate, and insist upon native medicines, their famous lotion being a decoction of Mhowa blossoms, which in my opinion aggravated the inflammation of the wound.

I returned Bisgaum to the Commissariat stables at Jubbulpur directly that he could march, as he was too uncontrollable for sporting purposes. Had any person been upon his back during his stampede he would have been swept off by the branches and killed; the mahout, sitting low upon his neck, could accommodate his body to avoid the boughs.

The use of the elephant in India is so closely associated with tiger-shooting that I shall commence the next chapter with the tiger.

CHAPTER V

THE TIGER

THERE is no animal that has exercised the imagination of mankind to the same degree as the tiger. It has been the personification of ferocity and unsparing cruelty.

In Indian life the tiger is so closely associated with the elephant (as the latter is used in pursuit) that I select this animal in sequence to the former, from which in the ideas of sporting Indians it is almost inseparable.

It is necessary to commence the description of the tiger with its birth.
The female rarely produces more than three, and generally only two.
These arrive at maturity in about two years.

There is a considerable difference in the size of the male and female. I have both measured and weighed tigers, and I have found a great difference in their proportions, such as may be seen not only in many varieties of animals, but also in human beings; it is therefore difficult to decide upon the actual average tiger, as they vary in separate localities, according to the quantity of wild animals in the jungles which constitute their food. If the tiger has been born in jungles abounding with wild pigs and other animals, he will have been well fed since the day of his birth, therefore he will be a well-developed animal.

A well-grown tigress may weigh an average of 240 lbs. live weight. A very fine tiger will weigh 440 lbs., but if very fat, the same tiger would weigh 500 lbs. I have no doubt there may be tigers that exceed this by 50 lbs., but I speak according to my experience.

The length of a tiger will depend upon the system of measurement. I always carry a tape with me, and I measure them before they are skinned, by laying the animal upon the ground in a straight line, and not allowing it to be stretched by pulling at the head or tail, but taking it naturally as it lies, measuring from nose to tip of tail. I have found that a tiger of 9 feet 8 inches is about 2 inches above the average. The same tiger may be stretched to measure 10 feet.

No person who examines skins only can form any idea of the true proportions of a tiger. The hide, when stripped from a tiger of 9 feet 7 inches, weighs 45 lbs. if the animal is bulky. The head, skinned, weighs 25 lbs. These weights are taken from an animal which weighed 437 lbs. exclusive of the lost blood, which was quite a gallon, estimated at 10 lbs. This would have brought the weight to 447 lbs. The hide of this tiger, which measured 9 feet 7 inches when upon the animal, was 11 feet 4 inches in length when cured. I have measured many tigers, and the skins are always stretched to a ridiculous length during the process of curing; these would utterly mislead any naturalist who had not practical experience of the live animal.

The tiger of zoological gardens is a long lithe creature with little flesh, and, from the lack of exercise, the muscles are badly developed. Such a specimen affords a poor example of the grand animal in its native jungles, whose muscles are almost ponderous in their development from the continual exertion in nightly rambles over long distances, and in mortal struggles when wrestling with its prey. A well-fed tiger is by no means a slim figure, but on the contrary it is exceedingly bulky, broad in the shoulders, back, and loins, with an extraordinary girth of limbs, especially in the forearm and wrist. The muscles are tough and hard, and there are two peculiar bones unattached to the skeleton frame; these are situated in the flesh of either shoulder, apparently to afford extra cohesion of the parts, resulting in additional strength when striking a blow or wrestling with a heavy animal.

There is a great difference in the habits of tigers; some exist upon the game of the jungles, others prey specially upon the flocks and herds belonging to the villagers; the latter are generally exceedingly heavy and fat. A few are designated "man-eaters"; these are sometimes naturally ferocious, and having attacked a human being, they may have devoured the body and thus have acquired a taste for human flesh; or they may have been wounded upon more than one occasion and have learnt to regard man as a natural enemy; but more frequently the man-eater is a wary old tiger, or more probably a tigress, that, having haunted the neighbourhood of villages and carried off some unfortunate woman when gathering firewood or the wild products of the jungles, has discovered that it is far easier to kill a native than to hunt for the scarce jungle game; the animal therefore adopts the pursuit of man, and seldom attempts to molest the natives' cattle.

A professed man-eater is the most wary of animals, and is very difficult to kill, not because it is superior in strength, but through its extreme caution and cunning, which renders its discovery a work of long labour and patient search. An average native does not form a very hearty meal. If a woman, she will have more flesh than a man about the buttocks, which is the portion both in animals and human beings which the tiger first devours. The maneater will seize an unsuspecting person by the neck, and will then drag the body to some retreat in which it can devour its prey in undisturbed security. Having consumed the hind-quarters, thighs, and the more fleshy portions, it will probably leave the body, and will never return again to the carcase, but will seek a fresh victim, perhaps at some miles' distance, in the neighbourhood of another village. Their cautious habits render it almost impossible to destroy a cunning man-eater, as it avoids all means of detection. In this peculiarity the ordinary man-eating tiger differs from all others, as the cattle-killer is almost certain to return on the following night to the body which it only partially devoured after the first attack. If the hunter has the taste and patience for night shooting, he will construct a hiding-place within 10 yards of the dead body. This should be arranged before noon, in order that no noise should disturb the vicinity towards evening, when the tiger may be expected to return. A tree is not a favourable stand for night shooting, as the foliage overhead darkens the sight of the rifle. Three poles of about 5 inches diameter and 12 feet in length should be sunk as a triangle, the thickest ends placed 2 feet in the ground. The poles should be 4 feet apart, and when firmly inserted will represent a scaffolding 10 feet high. Bars and diagonal pieces must be firmly lashed to prevent the structure from swaying. Within a foot of the top three strong cross-bars will be lashed, to support a corduroy arrangement of perfectly straight level bars, quite close together to form a platform. A thickly folded rug will carpet the rough surface, upon which the watcher will sit upon a low turnstool that will enable him to rest in comfort, and turn without noise in any required direction. A bamboo or other straight stick will be secured as a rail around the platform, upon which some branches may be so arranged as to form a screen that will conceal the watcher from the view of an approaching tiger. This arrangement is called a "mucharn."

When a tiger is driven before beaters it seldom or never looks upwards, but merely regards the surface as it advances; but when approaching a "kill" (the term applied to the animal which has been killed) the tiger is exceedingly cautious, and surveys everything connected with the locality before it ventures to recommence the feast. Even then, when assured of safety, it seldom eats the carcase where it lies, but seizing it by the throat, it drags the prey some 15 or 20 yards from the spot before it indulges in the meal. I have already described that the first meal consists of the buttocks and hindquarters; the second visit is devoted to the forequarters, after which but little remains for the vultures and jackals.

It is essential that the night watcher should be raised about 10 feet above the ground, otherwise the tiger would probably obtain his scent.

Night shooting is not attractive to myself, and I very seldom have indulged in such wearisome shikar. There is no particular satisfaction in sitting for hours in a cramped position, with mosquitoes stinging you from all directions, while your eyes are straining through the darkness, transforming every shadow into the expected game. Even should it appear, unless the moon is bright you will scarcely define the animal. I have heard well-authenticated accounts of persons who have patiently watched until they fell asleep from sheer weariness, and when they awoke, the dead bullock was no longer there, the tiger having dragged it away without disturbing the tired watcher. There are several methods of rendering the muzzle-sights of the rifle visible in partial darkness. A simple and effective arrangement is by a piece of thick white paper. This should be cut into a point and fastened upon the barrel with a piece of beeswax or shoemaker's wax, in addition to being tied with strong waxed packthread.

If a bright starlight night and there is no foliage above the rifle, the white paper will be distinctly seen, especially if the light is behind the shoulder. A piece of lime made into thick paste, and stuck upon the muzzle-sight, is frequently used by native hunters; but if it is at hand, there is nothing so effective as luminous paint; this can be purchased in stoppered bottles and will last for years. A small supply would be always useful in an outfit.

A man-eating tiger requires peculiar caution, not only lest it should observe the presence of the hunter, but he must remember that if upon the ground he himself becomes a bait for this exceedingly stealthy animal, which can approach without the slightest noise, and attack without giving any notice of its presence. A curious example of this danger was given a few years ago in the Nagpur district. A tigress had killed so many people that a large reward was offered for her destruction; she had killed and dragged away a native, but being disturbed, she had left the body without eating any portion. The shikaris considered that she would probably return to her prey during the night, if left undisturbed upon the spot where she had forsaken it. There were no trees, nor any timber that was suitable for the construction of a mucharn; it was accordingly resolved that four deep holes should be dug, forming the corners of a square, the body lying in the centre. Each hole was to be occupied by a shikari with his matchlock. The watchers took their positions. Nothing came; until at length the moon went down, and the night was dark. The men were afraid to get out of their hiding-places to walk home through the jungles that were infested by the man-eater; they remained in their holes, and some of them fell asleep.

When daylight broke, three of the shikaris issued from their positions, but the fourth had disappeared; his hole was empty! A few yards distant, his matchlock was discovered lying upon the ground, and upon the dusty surface were the tracks of the tiger, and the sweeping trace where the body had been dragged as the man-eater carried it along. Upon following up the track, the remains of the unlucky shikari were discovered, a considerable portion having been devoured; but the tigress had disappeared. This cunning brute had won the game, and she was not killed until twelve months afterwards, although many persons devoted themselves to her pursuit.

Many incredible stories have been told concerning the power of a tiger in CARRYING away his prey, and I have heard it positively stated by persons who should have known better, that a tiger can carry off a native cow simply through the strength of the jaws and neck. This is ridiculous, as the height of the cow exceeds that of the tiger, therefore a portion of the body must drag upon the ground. The cattle of India are exceedingly small, and are generally lean, the weight of an ordinary cow would hardly exceed 350 or 400 lbs.; as an average male tiger weighs about the same, it can of course drag its own weight by lifting the body partially in its mouth, and thus relieving the friction upon the ground. In this manner it is astonishing to see the strength exerted in pulling and lifting a dead bullock over projecting roots of trees, rocky torrent-beds, and obstructions that would appear to be insurmountable; but it is absurd to suppose that a tiger can actually lift and carry a full-grown cow or bullock in its jaws without leaving a trace of the drag upon the surface.

Many persons when in pursuit of tigers are accustomed to tie up a small buffalo of four or six months old for bait; the natives will naturally supply the poorest specimen of their herds, unless it is specially selected; therefore it may be quite possible for a large male tiger to carry so small an animal without allowing any portion of the body (excepting the legs) to drag upon the ground. As a rule, the tiger will not attempt to carry, but it will lift and pull simultaneously if the body is heavy.

The attack of a large tiger is terrific, and the effect may be well imagined of an animal of such vast muscular proportions, weighing between 400 and 500 lbs., springing with great velocity, and exerting its momentum at the instant that it seizes a bullock by the neck. It is supposed by the natives that the tiger, when well fastened upon the crest, by fixing its teeth in the back of the neck at the first onset, continues its spring so as to pass over the animal attacked. This wrenches the neck suddenly round, and as the animal struggles, the dislocation is easily effected. The tiger then changes the hold to underneath the throat, and drags the body to some convenient retreat, where the meal may be commenced in security. With very few exceptions the tiger breaks the neck of every animal it kills. Some persons have imagined that this is done by a blow of the paw, but this is an error. The tiger does not usually strike (like the lion), but it merely seizes with its claws, and uses them to clutch firm hold, and to lacerate its victim. I have seen several examples of the tiger's attack upon man, and in no instance has the individual suffered from the shock of any blow; the tiger has seized, and driven deeply its claws into the flesh, and with this tremendous purchase it has held the victim, precisely as the hands of a man would clutch a prisoner; at the same time it has taken a firm hold with its teeth, and either killed its victim by a crunch of the jaws, or broken the shoulder-blade. In attacking man the tiger generally claws the head, and at the same moment it fixes its teeth upon the shoulder. An Indian is generally slight, and shallow in the chest, therefore the wide-spread jaws can include both chest and back when seized in the tiger's mouth. I have seen men who were thus attacked, and each claw has cut down to the skull, leaving clean incisions from the brow across the forehead and over the scalp, terminating at the back of the neck. These cuts were as neatly drawn across the skull as though done by a sharp pruning-knife; but the wounded men recovered from the clawing; the fatal wound was the bite, which through the back and chest penetrated to the lungs.

It is surprising that so few casualties occur when we consider the risks that are run by unprotected natives wandering at all seasons through the jungles, or occupied in their daily pursuits, exposed to the attacks of wild animals. The truth is that the tiger seldom attacks to actually kill, unless it is driven, or wounded in a hunt. It will frequently charge with a short roar if suddenly disturbed, but it does not intend to charge home, and a shout from a native will be sufficient to turn it aside; it will then dash forward and disappear, probably as glad to lose sight of the man as he is at his escape from danger. Of course there are many exceptions when naturally savage tigers, without being man-eaters, attack and destroy unoffending natives without the slightest provocation; upon such occasions they leave the body uneaten, neither do they return to it again.

Although the tiger belongs to the genus Felis, it differs from the cat in its peculiar fondness for water. In the hot season the animal is easily discovered, as it invariably haunts the banks of rivers, when all the brooks are dry and the tanks have disappeared through evaporation. The tiger loves to wallow in shallow water, and to roll upon the dry sand after a muddy bath; it will swim large rivers, and in the Brahmaputra, where reedy and grassy islands interrupt the channel in a bed of several miles' width, the tigers travel over considerable distances during the night, swimming from island to island, and returning to the mainland if no prey is to be found during the night's ramble.

The tiger is by no means fond of extreme heat; it is found in northern China, Manchuria, and the Corea, where the winters are severe. In those climates during winter the skin is very beautiful, consisting of thick fur instead of hair, and the tail is comparatively bushy. Well-preserved skins of that variety are worth 20 pounds apiece and are prized as rarities. In the hot season of India the tiger is by no means happy: it is a thirsty animal, and being nocturnal, it quickly becomes fatigued by the sun's heat, and the burning surface of the soil if obliged to retreat before a line of beaters. The pads of the feet are scorched by treading upon heated sandy or stony ground, and the animal is easily managed in a beat by those who are thoroughly experienced in its habits, although during the winter season, when water is abundant in all the numerous nullahs and pools, there is no animal more difficult to discover than the tiger. It may be easily imagined that the dense green foliage of Indian jungles renders all objects difficult to perceive distinctly, but the striped skin of a tiger harmonizes in a peculiar manner with dry sticks, yellowish tufts of grass, and the remains of burnt stumps, which are so frequently the family of colours that form the surroundings of the animal. In this covert the tiger with an almost noiseless tread can approach or retreat, and be actually within a few yards of man without being seen. Although a ferocious beast, it is most sensitive to danger, and the slightest noise will induce it to alter the direction of its course when driven before a line of beaters. Its power of scent is excellent, therefore it is always advisable if possible to arrange that the beaters shall advance down wind. If they do, the tiger may be generally managed so adroitly that it will be driven in the required direction; but if the beaters are travelling up the wind, the tiger must necessarily follow the same course, and it will probably obtain the scent of the guns that are in positions to intercept it, in which case it will assuredly dash back through the line of beaters, and escape from the beat.

In the hot season very few trees retain their leaves, and the jungles that were impervious screens during the cooler months become absolutely naked; an animal can then be discerned at 100 yards' distance. The surface of the ground is then covered with dried and withered leaves, which have become so crisp from the extreme heat that they crackle when trod upon like broken glass. It will be readily understood that any form of shooting excepting driving is quite impossible under these conditions, as no person could approach any animal on foot owing to the noise occasioned by treading upon the withered leaves.

The habits of the tiger being thoroughly understood, it becomes necessary under all circumstances to employ the village shikari. This man is generally more or less ignorant and obstinate, but he is sure to know his own locality and the peculiar customs of the local tiger. It is one of the mysterious characteristics of this animal that it invariably selects particular spots in which it will lay up; to these secure retreats it will retire; therefore, should a fresh track be discovered upon the sandy bed of a nullah or upon a dusty footpath in the jungles, it may be safely inferred that the tiger is lying in one or other of its accustomed haunts. The village shikari will quickly determine from what direction the tiger has arrived; he will then suggest the probable route that the animal will take whenever it may be disturbed.

Should the tiger be killed, another will occupy its place a few months later, and this will assuredly assume the same habits as its predecessor; it will frequent the same haunts, lay up in the same spots, and drink at the same places; although it may have never associated with or even seen the tiger which formerly occupied the same locality.

I have already described the keen power of scent possessed by this wary animal, which necessitates extreme caution, and the placing of the guns in positions elevated about 10 feet above the ground. It is seldom of any use to drive jungles upon speculation, although it not unfrequently happens, where tigers are plentiful, that when driving for deer the grander game unexpectedly appears, and presents itself suddenly before the astonished hunter. The recognised system of tiger-hunting by driving is as follows. We will say that the party of three may have arrived at a village, after having received intimation that a native cow had been carried off within the last few days. The first operation is to send natives in all directions to look for tracks, and to discover the place where the animal last drank.

At least two elephants should accompany the party, even though the thick jungle country may be ill adapted for shooting from these useful creatures. One of these should be, if possible, a really dependable animal, that would advance steadily and quietly up to a wounded tiger. The great danger of this branch of sport arrives when a tiger may have been wounded, and it has to be tracked up on foot, and eventually beaten out of the dense thorny cover of its retreat. A staunch elephant is then indispensable, and the real excitement commences when the beaters are sent for safety up the adjoining trees, and the hunter, absolutely certain that the dangerous game, although invisible, is close before him, advances calmly to the attack, knowing that the tiger will be ready to spring upon the elephant the moment that they shall be vis-a-vis.

In the absence of any elephant, the pursuit of a wounded tiger by following up the blood-track on foot is a work of extreme danger. The native shikaris generally exhibit considerable hardihood, and, confident in their activity, they ascend trees from which they have a clear view in front for some 30 or 40 yards. They descend if the coast is clear, cautiously advance, and then again they mount upon the branches of some favourable tree and scan the ground before them. In this manner they continue to approach until they at length discern the wounded animal. If the hunter is clever at climbing, he may then take a steady shot from a good elevation; but if not, he must take his chance, and knowing the exact position of the tiger, he must endeavour to make certain of its sudden death by placing a bullet either in the brain or the back of the neck.

A newly arrived party, having heard that some native cow has been carried off within a week, will make a reconnaissance of the surrounding country upon their elephants, and will examine every watercourse for tracks. We will suppose that after some hours of diligent search the long-wished-for pugs or footmarks have been discovered. Now the science of the chase must be exhibited, and the habits of the tiger carefully considered. The first consideration will be the drinking-place. If the middle of the dry season, say the beginning of May, the heat will be intense, and the hot wind will feel as though it had passed over a heated brick-kiln. The water will have entirely disappeared, unless a river shall be permanent in the neighbourhood. It will be necessary to procure two or perhaps three buffaloes to tie up in various positions not far from water, as baits for the tiger during the hours of night, when it will be wandering forth from its secure retreat and searching for its expected prey. The buffaloes should be at least twelve months old; I prefer them when eighteen months, as they are then heavy animals and would afford two hearty meals, each sufficient to gorge the tiger to an extent that, after drinking, would render it lazy and inclined to sleep. Great care should be taken in the selection of these buffaloes. The natives will assuredly offer their skinny and unhealthy animals: but a tiger, unless nearly starved, will frequently refuse to attack a miserable skeleton, and like ourselves it prefers a fat and appetising attraction. It must be distinctly remembered that after the tiger has devoured the hind-quarters of the animal it has killed, it requires a deep draught of water; it is therefore necessary that the buffalo as bait should be tied up somewhere within a couple of hundred yards of a drinking-place, as the least distance; otherwise, instead of lying down somewhere near the remains of its prey, it must wander to a great distance to drink. The stomach, being full of flesh, will naturally become distended with water, and the gorged tiger will not be in the humour to undertake a return journey of perhaps a mile to watch over the remains of its kill; it will therefore lie down in some thick covert near the spot by the nullah where it recently drank, instead of returning to repose in the neighbourhood of its recent victim. This will throw out the calculations of the shikari, who would expect that the tiger will be lying somewhere near the spot where it dragged the buffalo. The beat will under such false conditions be arranged to include an area in which the tiger is supposed to be asleep after its great meal, but in reality it may be a mile or two away in some unknown direction near the water. Great precaution is necessary in making all preliminary arrangements. It is a common custom of native shikaris to tie up a buffalo where four paths meet, as the tiger would be walking along one of these during the night, and it could not help seeing the alluring bait. I do not admire this plan, as, although the probability is that the buffalo will be killed, there is every likelihood of disturbance after the event, when natives would be passing along the various routes. The slightest noise would alarm the tiger, and instead of remaining quietly near the carcase, it would slink away and be no more seen.

Natives are very inquisitive, and should the tiger have killed the bait, and dragged the buffalo away to some deep nullah, the shikari and his companion are often tempted to creep along the trace until they perhaps see the tiger in the act of devouring the hind-quarters. This is quite contrary to the rules of hunting, as the tiger is almost certain to detect their presence if they are so near, in which case it is sure to retreat to some undisturbed locality beyond the area of the beat.

There is constant disappointment in driving for tigers owing to the stupidity or exaggerated zeal of the shikari; and if the hunter is thoroughly experienced, it is far better that he should conduct the operations personally.

Success depends upon many little details which may appear trivial, but are nevertheless important. When a buffalo is tied up for bait, it must be secured by the fetlock of a fore foot, and care must be taken that the rope is sufficiently strong to prevent the buffalo from breaking away; at the same time it must not be strong enough to prevent the tiger from breaking it when the animal is killed, and the carcase is to be dragged to the nearest nullah (or ravine). If the rope is too powerful, the tiger cannot dispose of the body; it will therefore eat the hind-quarters where it lies, and at once retreat to water, instead of concealing the prey and lying down in the vicinity. In such a case the remains of the body will be exposed to the gaze of vultures and jackals, who will pick the bones clean in a few hours, and destroy all chance of the tiger's return. When the dead body is concealed beneath dense bushes in a deep ravine, the vultures cannot discover it, as they hunt by sight, and the tiger has no anxiety respecting the security of its capture; it will therefore sleep in peace within a short distance, until awakened by the shouts of a line of beaters.

If the buffalo is tied with a rope around the neck, a tiger will frequently refuse to molest it, as it fears a trap. I have seen occasions when the tiger has walked round and round the buffalo, as exhibited by the tracks upon the surface, but it has been afraid to make its spring, being apprehensive of some hidden danger. I have also seen a dead vulture lying close to the body of a buffalo, evidently killed by a blow from the tiger's paw when trespassing upon the feast. It is a good arrangement to secure both fetlocks of a buffalo with a piece of strong cord about a foot or 16 inches apart, independently of the weaker cord which ties the animal to either a stake or tree. Should the buffalo break away during the night, it cannot wander far, as the bushes will quickly anchor the rope which confines the fore legs; the tiger would then assuredly attack the straying animal and kill it within the jungles. In such a case the drive should take place without delay, as the dead buffalo will certainly be hidden in the nearest convenient spot, and the tiger will be somewhere in the neighbourhood.

During the hot season it will be advisable to defer the drive till about IO A.M., at which time the tiger will be asleep. The mucharns or watching-places in various trees should have been previously constructed before the buffaloes were tied up in their different positions, to be ready should the tiger kill one of the baits, and thus to avoid noise during the construction. This is a matter of very great importance which is frequently neglected by the native shikari, who postpones the building of mucharns until the tiger shall have killed a buffalo. In that case the noise of axes employed in chopping the wood necessary for building the platforms is almost sure to alarm the tiger, who will escape unseen, and the beat will take place in vain.

I never allow mucharns to be built by wood felled in the immediate neighbourhood, but I have it prepared in camp, and transported by coolies to the localities when required. By this method the greatest silence may be observed, which is absolutely necessary to ensure a successful drive.

In order to prepare these platforms, they should be laid upon the ground, three long thick pieces to form a triangle, and cross-bars in proportionate lengths. If the latter are straight and strong, from sixteen to twenty will be necessary to complete a strong mucharn. It is impossible to devote too much attention to the construction of these watching-places. The natives are so light, and they are so comfortable when squatting for hours in a position that would cramp a European, that it is dangerous to accept the shikari's declaration when he reports that everything is properly arranged. Upon many occasions tigers are missed because the shooter is so completely cramped that he cannot turn when the animal suddenly appears in view. A large, firm, and roomy mucharn fixed upon the boughs of a tree that will not wave before a gust of wind, is the proper platform to ensure a successful shot.

I have frequently been perched in a mere heron's nest, formed of light wood arranged upon most fragile boughs; this wretched contrivance has swayed before the wind to an extent that would have rendered accurate aim impossible; fortunately upon such occasions I have never obtained a shot.

Although driving may read as an unexciting sport, it is quite the contrary if the hunter takes sufficient interest in the operations to attend to every detail personally. When all is in readiness after the tiger has killed a buffalo, there is much art required in the conduct of the drive. Natives vary in different districts; some are clever and intelligent, and take an immense interest in the sport, especially if they are confident in the generosity of their employer. In other districts there may be abundant game, but the natives are cowardly, and nothing will persuade them to keep an unbroken line, upon the perfection of which the success of the drive depends.

As a rule, there is no great danger in the steady advance of a line of men, provided they are at close intervals of 5 or 8 yards apart, and that they keep this line intact. It is a common trick, when the beaters are nervous, to open out the line in gaps, and the men resolve themselves into parties of ten or twenty, advancing in knots, at the same time howling and shouting their loudest to keep up the appearance of a perfect line. In such cases the tiger is certain to break back through one of the inviting gaps, and the drive is wasted.

To drive successfully, the beaters must not only keep a rigid line, but they must thoroughly understand the habits of the animal, and the positions of the posted guns. If the drive is thoroughly well organised, there should be eight or ten men who are experienced in the sport; these should take the management of the beat, and being distributed at intervals along the line, they should direct the operations.

A few really clever shikaris should be able (with few exceptions to the rule) to drive the tiger to any required position, so as to bring it within shot of any particular mucharn. This may be effected without extraordinary difficulty. The drive should be arranged to include three parts of a circle. If there are three guns, their positions would depend upon the quality and conditions of the ground, leaving intervals of only 80 or 100 yards at farthest between the three mucharns. From either flank, commencing only 50 yards from each mucharn, a native should be posted in a tree, and this system of watchers should be continued until they meet the extreme ends of the right and left flanks of the beating line. It will be seen that by this method there is a chain of communication established throughout the line, both flanks being in touch with the right and left mucharns by watchers in the trees only 50 yards apart. The tiger, if within the beat, will be completely encircled, as it will have the guns in front, the line of beaters in a semicircle behind, and a chain of watchers in trees from 30 to 50 yards apart from either side of the line to within sight of the mucharns. If the jungle should be tolerably open, the tiger cannot move without being seen by somebody. It now has to be driven before the beaters, and it should be induced to select a particular direction that will bring it within distance of one particular mucharn.

Each man who may be perched in the trees, which form a chain from the right and left extremities of the line, will be provided with several pieces of exceedingly dry and brittle sticks; he will hold these in readiness for use whenever he may observe the tiger. If he sees that the animal wishes to pass through the line, and thereby escape from the beat, he simply breaks a small stick in half; the sound of a snap is quite sufficient to divert the tiger from its course; it will generally stop and listen for a few moments, and then being alarmed by the unusual sound, it will again move forward, this time in the required direction, towards the guns. In this manner the animal is gradually guided by the unseen watchers in the trees, and is kept under due control, without any suspicion upon its part that it is being conducted to the fatal spot within 30 or 40 yards of the deadly aim of an experienced rifle. This leading of the tiger requires considerable skill, as much discretion is necessary in breaking the stick at the proper moment, or increasing the noise should it be deemed expedient.

As a rule, the slightest sound is sufficient to attract the attention of a driven tiger, as the animal is well aware that the shouts of a line of beaters are intended to scare it from the neighbourhood; it is accordingly in high excitement, and it advances like a sly fox slowly and cautiously, occasionally stopping, and turning its head to listen to the cries of the approaching enemy. Any loud and sudden noise would induce it to turn and charge back towards the rear, in which case it is almost certain to escape from the beat.

Some tigers are more clever than others, and having escaped upon more than one occasion, they will repeat the dodge that has hitherto succeeded. It is a common trick, should the jungle be dense and the ground much broken, for the tiger to crouch when it hears the beaters in the distance, instead of going forward in the direction of the guns. This is a dangerous stratagem, as the wary animal will lie quietly listening to the approaching line, and having waited until the beaters are within a few yards of its unexpected lair, it will charge back suddenly with a terrific roar, and dash at great speed through the affrighted men, perhaps seizing some unfortunate who may be directly in its path. I have known tigers that have been hunted many times, but who have always escaped by this peculiar dodge, and such animals are exceedingly difficult to kill. In such cases I am of opinion that no shouts or yells should be permitted, but that the line should advance, simply beating the stems of trees with their sticks; at the same time six or eight natives with their matchlocks should be placed at intervals along the line to fire at the tiger should it attempt to break through the rear. This may sometimes, but rarely, succeed in turning it, and compelling it to move in the required direction. It is a curious fact that "breaking back" is a movement general to all animals, which have an instinctive presentiment of danger in the front, if alarmed by the sound of beaters from behind. If once they determine upon a stampede to the rear, nothing will stop them, but they will rush to destruction and face any opposition rather than move forward before the line. The tiger in such cases is extremely dangerous, although when retreating in an ordinary manner before the beaters it would seldom attack a human being, but, on the contrary it would endeavour to avoid him. It is frequently the custom of tigers to remain together in a family the male, female, and a couple of half or three parts grown young ones. We cannot positively determine whether the male always remains with his family under such circumstances, or whether he merely visits them periodically; I am inclined to the latter opinion, as I think the female may be attractive during her season, which induces the male to prolong his visit, although at other periods he may be leading an independent life. Good fortune specially attends some favoured sportsmen who have experienced the intensity of happiness when a complete family of tigers has marched past their position in a drive, and they have bagged every individual member. This luck has never waited upon me, but I have seen three out of the four secured, the big and wary male, having modestly remained behind, escaping by breaking back through the line of beaters.

The tigress remains with her young until they are nearly full-grown, and she is very assiduous in teaching her cubs to kill their prey while they are extremely young. I have seen an instance of such schooling when two buffaloes were tied up about a quarter of a mile apart; one was killed, and although these two baits were mere calves, it had evidently been mangled about the neck and throat in the endeavour to break the neck. This had at length been effected by the tigress, as proved by the larger marks of teeth, while the wounds of smaller teeth and claws in the throat and back of neck showed that the cub had been worrying the buffalo fruitlessly, until the mother had interfered to complete the kill. The other buffalo calf had been attacked, and severely lacerated about the nape of the neck and throat, but it was still alive, and was standing up at the post to which it had been tied. This proved that the cub had been practising upon both these unlucky animals, and that the tigress had only interfered to instruct her pupil upon the last occasion. A dead vulture was lying near the buffalo carcase; this had been killed, probably, by the cub; the fact showed that the buffalo had been attacked that morning during daylight, and not during the preceding night, when the vultures would have been at roost.

The tigress is generally in advance of the male during a drive, should there be two together; this should not be forgotten, and a sharp look-out should be directed upon the place from whence the tigress shall have emerged, as the shot must be taken at the rearmost animal, who would otherwise disappear immediately, and break back at the sound of the explosion. In all cases it is incumbent upon the watcher to study attentively every feature of the ground directly that he enters upon his post, so that he may be prepared for every eventuality; he should thoroughly examine his surroundings, noting every little open space, every portion of dense bush, and form his opinion of the spot that would probably be the place of exit when the tiger should be driven to the margin of the covert. Tigers are frequently missed, or only slightly wounded, through utter carelessness in keeping a vigilant look-out. The watcher may have omitted to scan the details of the locality, and when unprepared for the interview, the tiger suddenly appears before him. Startled at the unexpected apparition, he fires too quickly, and with one bound the tiger vanishes from view, leaving the shooter in a state of misery at his miss, that may be imagined. Nearly all the fatalities in tiger-shooting are caused by careless shooting, which necessitates the following up a blood-track; it is therefore imperative that extreme care and coolness be observed in taking a steady aim at a vital portion of the body, that will ensure the death of the animal at latest within a few minutes. If the shot is fired at right angles with the flank, exactly through the centre of the blade-bone, the tiger will fall dead, as the heart will be shattered, and both shoulders will be broken. A shot close behind the shoulder will pass through the centre of the lungs, and death will be certain in about two minutes, but the animal will be able to inflict fatal injuries upon any person it may encounter during the first minute, before internal bleeding shall have produced complete suffocation. If the hunter is confident in the extreme accuracy of his rifle, a shot in the centre of the forehead rather above a line drawn across the eyes will ensure instant death. This is a splendid shot when the hunter sits upon an elevation and the tiger is approaching him; in that position he must be careful to aim rather high, as, should the bullet miss the forehead, it will then strike the spine at the junction of the neck; or if too high, it will break the spine between the shoulders; at any rate, the chances are all in favour of the rifle, whereas, should the aim be too low, the bullet might penetrate through the nose, and bury itself within the ground, merely wounding the animal instead of killing. Should the hunter be on foot, he must on the contrary aim low, exactly at the centre of the nose; if he is only one inch too high, the tiger may escape, as the bullet may pass over the head and back; but if the aim is low and the nose should be missed, the bullet will either break the neck, or regularly rake the animal by tearing its course through the chest and destroying the vitals in its passage along the body. In that case the .577 solid bullet of 650 grains and 6 drams of powder will produce an astonishing effect, and will completely paralyse the attack of any lion or tiger, thus establishing a thorough confidence in the heart of its proprietor.

CHAPTER VI

THE TIGER (continued)

There is no more delightful study than Natural History in its practical form, where the wild beasts and their ways are actually presented to the observer in their native lands, and he can examine their habits in their daily haunts, and watch their characters in their wild state instead of the cramped limits of zoological collections. At the same time we must confess that the animals of a menagerie afford admirable opportunities for photography, and are most instructive for a rudimentary preparation before we venture upon the distant jungles where they are to be found in their undisturbed seclusion. It is commonly supposed that wild animals that have never been attacked by fire-arms are not afraid of man, and that deer, antelopes, and various species which are extremely timid may be easily approached by human beings, as the creatures have no fear of molestation. My experience does not support this theory. Nearly all animals have some natural enemy, which keeps them on the alert, and renders them suspicious of all strange objects and sounds that would denote the approach of danger. The beasts of prey are the terror of the weaker species, which cannot even assuage their thirst in the hottest season without halting upon the margin of the stream and scrutinising the country right and left before they dare stoop their heads to drink. Even then the herd will not drink together, but a portion will act as watchers, to give notice of an enemy should it be discerned while their comrades slake their thirst.

It is a curious and inexplicable fact that certain animals and varieties of birds exhibit a peculiar shyness of human beings, although they are exposed to the same conditions as others which are more bold. We see that in every portion of the world the curlew is difficult to approach, although it is rarely or never pursued by the natives of the neighbourhood; thus we find the same species of bird exhibiting a special character whether it has been exposed to attack, or if unmolested in wild swamps where the hand of man has never been raised against it.

The golden plover is another remarkable example, as the bird is wild in every country that it inhabits, even where the report of fire-arms never has been heard. The wagtails, on the contrary, are tame and confiding throughout all places, whether civilised or savage. The swallows are the companions of the human race, nesting beneath their eaves, and sharing the shelter of their roofs in every clime. Why this difference exists in creatures subjected to the same conditions is a puzzle that we cannot explain. In like manner we may observe the difference in animals, many of which are by nature extremely timid, while others of the same genus are more bold. The beasts of prey vary in an extraordinary degree according to their species, which are in some way influenced by circumstances. Tigers and lions are naturally shy, and hesitate to expose themselves unnecessarily to danger; both these animals will either crouch in dense covert and allow the passer-by to continue his course, or slink away unobserved, if they consider that their presence is undetected. Nevertheless these animals differ in varying localities, and it is impossible to describe the habits of one particular species in general terms, as much depends upon the peculiarities of a district which may exercise an effect in influencing character. The tigers that inhabit high grass jungle are more dangerous than those which are found in forests. The reason is obvious; the former cannot be seen, neither can they see, until the stranger is almost upon them; they have accordingly no time for consideration, but they act upon the first impulse, which is either to attack in self-defence or to bound off in an opposite direction. If the same tiger were in a forest it would either see the approach or it would hear the sound of danger, and being forewarned, it would have time to listen and to decide upon a course of retreat; it would probably slink away without being seen.

Although the usual bait for a tiger is a young buffalo, there is no animal that is held in greater respect by this ferocious beast than an old bull of that species.

It is by no means an uncommon occurrence that should a tiger have the audacity to attack a buffalo belonging to a herd, the friends of the victim will immediately rush to its assistance, and the attacking party is knocked over and completely discomfited, being only too glad to effect a retreat.

A few months ago, from the date at which I am now writing, a native came to my camp with the intelligence that a large tiger had suddenly sprung from a densely wooded nullah and seized a cow that was grazing within a few yards of him. The man shouted in the hope of scaring the tiger, when two buffaloes who were near the spot and were spectators of the event at once charged the tiger at full speed, knocked it over by their onset, and followed it as it sprang for safety into the thick bush, thus saving the cow from certain destruction. The cow, badly lacerated about the throat, ran towards its native village, followed by its owner. I lost no time in arriving at the spot, about two miles from camp, and there I found the recent tracks precisely tallying with the description I had received. We organised a drive on the following morning, but the crestfallen tiger had taken the notice to quit, and had retreated from the neighbourhood.

An example of this kind is sufficient to exhibit the cautious character of the tiger. My shikari, a man of long experience, differed in opinion with the native who had witnessed the attack. This man declared that the tiger must be lying in a dense thicket covering a deep hollow of about 10 acres, to which it had retreated when charged by the two buffaloes; he advised that we should lose no time, but organise a drive at once, as the tiger, having been frightened by the buffaloes, would probably depart from the locality during the night.

My shikari argued against this suggestion. He was of opinion that the tiger might not be lying in the hollow, as there was much broken ground and jungle in the immediate neighbourhood, including many dense and deep nullahs that might have formed a retreat: if the tiger should happen to be within one of those places, it would be outside the drive, and would be frightened away by the noise of the beaters should we drive the hollow, and it would escape unseen. If, on the other hand, the tiger should be lying in any spot within a radius of half a mile, it would be very hungry, as proved by its attack upon the cow during broad daylight, and it would assuredly kill one or both of the baits, and remain with its prey, if we should tie up two young buffaloes that night; we should then be certain to have it within the drive on the following morning.

This was sound reasoning, and according to rule; but the native argued that the tiger, having been knocked over and pounded by the buffaloes, would be so cowed that it would decline to attack the young buffaloes that might be secured to trees as baits; it would, on the contrary, avoid anything in the shape of a buffalo, and if we neglected to drive the jungle at once, we should find a blank upon the following morning.

The sequel proved that the man was correct, as the buffaloes were untouched on the following day, and the tiger had disappeared from the locality.

The tiger, although hungry, was sufficiently disturbed by its defeat to abstain from any further attack; although the baits were only twelve months old, it was too shy to encounter anything in the shape of a buffalo.

In the grassy islands of the Brahmaputra there were a vast number of tigers some twelve or fourteen years ago, but their number has been reduced through the development of the country by the various lines of steamers which have improved the navigation of the river. Formerly a multitude of small islands of alluvial deposit thrown up by the impetuous current created an archipelago for 60 or 70 miles of the river's course south of Dhubri, in the direction of Mymensing; these varied in size from a few hundred yards to a couple of miles in length, and being covered with high grass and tamarisk, they formed a secluded retreat for tigers and other game at the foot of the Garo Hills. The river makes a sudden bend, sweeping near the base of this forest-covered range, from which the wild animals at certain seasons were attracted to the island pasturage and dense covert, especially when the forests had been cleaned by annual firing, and neither food nor place of refuge could be found. As these numerous islands abounded with wild pigs, hog-deer, and other varieties of game, they were most attractive to tigers, and these animals were tolerably secure from molestation, as it was impossible to shoot or even to discover them in grass 10 feet high without a line of elephants. The improvement introduced by steam navigation gave an increased impulse to cultivation, as the productions of the country could be transported at a cheap rate to Calcutta by the large barges termed flats, which are fastened upon either side of the river steamers. These are 270 feet in length, and of great beam. The steamers are from 270 to 300 feet from stem to stern, and are furnished with hurricane decks capable of stowing a large cargo, although the draught of water is limited owing to the numerous sandbanks that interrupt the channel. The peculiar conditions of the Brahmaputra, which render it necessary that these large vessels should be of very shallow draught, entail the necessity of a rudder 17 feet in length to afford a sufficient resistance for steering when running down the stream. The shock when striking upon a sand-bank is sufficient to bury the stem without straining the vessel, as the flat bottom remains fixed upon the soft soil for a few moments, during which the force of the stream upon so large a surface brings the steamer broadside on to the obstruction and releases the stem. It is then an affair of an hour or more to get her off the bank by laying out kedge anchors, and heaving upon the hawsers with the steam winches.

The Brahmaputra is an extraordinary river, as it acknowledges no permanent channel, but is constantly indulging in vagaries during the season of flood; at such times it carries away extensive islands and deposits them elsewhere. Sometimes it overflows its banks and cuts an entirely new channel at a sudden bend, conveying the soil to another spot, and throwing up an important island where formerly the vessels navigated in deep water. This peculiar character of the stream renders the navigation extremely difficult, as the bed is continually changing and the captains of the steamers require a long experience.

During inundations the islands are frequently drowned out, and the wild animals are forced to swim for the nearest shore. Upon such occasions tigers have been frequently seen swimming for their lives, and they have been killed in the water by following them in boats. The captain of the steamer in which I travelled told me of a curious incident during a great inundation, which had covered deeply all the islands and transported many into new positions. Upon waking at daylight, the man who took the helm was astonished to see a large tiger sitting in a crouching attitude upon the rudder, which, as already explained, was 17 feet in length. A heavily-laden flat or barge was lashed upon either side, and the sterns of these vessels projected beyond the deck of the steamer, right and left.

The decks of these large flats were only feet above the water, and the tiger, when alarmed by a shout from the helmsman, made a leap from the rudder to the deck of the nearest vessel. In an instant all was confusion, the terrified natives fled in all directions before the tiger, which, having knocked over two men during its panic-stricken onset, bounded off the flat and sought security upon the deck of the steamer alongside. Scared by its new position and by the shouts of the people, it rushed into the first hole it could discover; this was the open door of the immense paddle-box, and the captain rushed to the spot and immediately closed the entrance, thereby boxing the tiger most completely.

There was only one gun on board, belonging to the captain: the door being well secured, there was no danger, and an ornamental air-hole in the paddle-box enabled him to obtain a good view of the tiger, who was sitting upon one of the floats. A shot through the head settled the exciting incident; and the men who were knocked over being more frightened than hurt, the affair was wound up satisfactorily to all parties except the tiger.

The progress of science in the improvement of steam navigation has had a wonderful effect throughout the world during the past half century, and it is interesting to watch the development resulting from the increased facilities of steam traffic upon the Brahmaputra. Although a residence upon the islands is accompanied by extreme risk during the period of inundations, there are many villages established where formerly the tigers held undisturbed possession; and the rich alluvial soil is made to produce abundance, including large quantities of jute, which is transported by the steamers to Calcutta. The danger of an unexpected rise in the river is always provided for, and every village possesses two or more large boats, which are carefully protected from the sun by a roof of mats or thatch, to be in readiness for any sudden emergency.

When the natives first established themselves upon the islands and along the dangerous banks of the Brahmaputra, they suffered greatly from the depredations of the numerous tigers, and in self-defence they organised a system by which each village paid a subscription towards the employment of professional shikaris. These men soon reduced the numbers of the common enemy, by setting clever traps, with bows and arrows, the latter having a broad barbed head, precisely resembling the broad arrow that is well known as the Government mark throughout Great Britain. The destruction of tigers was so great in a few years that the Lieut.-Governor of Bengal found it necessary to reduce the reward from fifty rupees to twenty-five, and tiger-skins were periodically sold by auction at the Dhubri Kutcherry at from eight annas to one rupee each.

In this manner the development of agricultural industry brought into value the fertile soil, which had hitherto been neglected, and the wild beasts were the first to suffer, and eventually to disappear from the scene; precisely as indolent savage races must vanish before the inevitable advance of civilisation. and their neglected countries will be absorbed in the progressive extension of colonial enterprise.

I believe there are very few tigers to be found at the present time in the islands or "churs" of the Brahmaputra, and although I never had the good fortune to know the country when it was described to me as "crawling" with these animals, I look back with some pleasure to my visit in 1885, when through the kindness of Mr. G. P. Sanderson, the superintendent of the keddahs, I was supplied with the necessary elephants.

The Rajah of Moochtagacha, Soochikhan (or Suchi Khan), had started from Mymensing with thirty-five elephants, and he kindly invited me to join him for a few days before I should meet Mr. Sanderson at Rohumari, about 38 miles below Dhubri, on the Brahmaputra. I had a scratch pack of twelve elephants, including some that had been sent forward from the keddahs, and others kindly lent by the Ranee of Bijni. These raised our number into a formidable line, excepting one huge male with long tusks belonging to the Bijni Ranee, who was too savage to be trusted with other elephants in company. This brute, as is not uncommon, combined great ferocity with extreme nervousness. He had just destroyed the howdah, which was smashed to atoms, as the animal had taken fright at the crackling of flames when some one had ignited a patch of long grass in the immediate neighbourhood. This had established an immediate panic, and the elephant bolted at full speed, destroying the howdah utterly beneath the branches of a tree; fortunately there was no occupant, or he would certainly have been killed. The sound of fire is most trying to the nerves of elephants, but a good shooting animal should be trained especially to bear with it; otherwise it is exceedingly dangerous.

The Rajah's elephants were his peculiar enjoyment, and there was the same difference in their general appearance, when compared with the keddah elephants, as would be seen in a well-kept stable of hunters and a team of ordinary farm horses. At the same time it must be remembered that Suchi Khan's elephants did no work, but were kept solely for his amusement, while the keddah animals had been working hard in the Garo Hills for many months upon inferior food, engaged with their experienced superintendent Mr. Sanderson in catching wild elephants. Nevertheless there was a notable superiority in the Rajah's shikari animals, as they had been carefully trained to the sport of tiger-hunting; they marched with so easy a motion that a person could stand upright in the howdah, rifle in hand, without the necessity of holding the rail. They appeared to glide instead of swaying as they moved, and in that respect alone they exhibited immense superiority, the difficulty of shooting with a rifle from the back of an elephant in motion being extreme. Several of these elephants were so well trained that they showed no alarm when a tiger was on foot, at which time an elephant generally exhibits a tendency to nervousness, and cannot be kept motionless by his mahout.

A favourite shikar animal had been badly bitten by a tiger a few days before my arrival, and it was feared that she might become shy upon the next encounter. Although the elephant is enormous in weight and strength, the upper portion of the trunk is much exposed, as it is the favourite spot for the tiger's attack, where it can fix its teeth and claws, holding on with great tenacity. A wound on the trunk is most painful, and when an elephant is actually pulled down by a tiger, it is the pain to which the animal yields in falling upon the knees, more than the actual weight and strength of the tiger that produce the effect. A tiger, when standing upon its hind legs, would be able to reach about 8 feet without the effort of a spring; it may be readily imagined that a female elephant unprotected by tusks must certainly be injured should a tiger rush determinedly to the attack; nevertheless the female is generally preferred to the male for steadiness and docility. When a really trustworthy male elephant is obtainable, well grown, of large size, easy action, and in perfect training, it is simply invaluable, and there is no pleasure equal to such a mount; the sensation upon such an animal is too delightful, and you long for the opportunity to exhibit the power and prowess of your elephant, as the feeling of being invincible is intensely agreeable. The only sensation that can approach it is the fact of being mounted upon a most perfect hunter, that you can absolutely depend upon when following the hounds in England; an animal well up to a couple of stones more than your own weight, who never bores upon your hand, but keeps straight, and never makes a mistake; even that only faintly approaches the pleasure of a good day upon such an elephant as I have described.

Mahouts will always lie concerning the reputation of the animal in their charge, and I had been assured that the great male belonging to the Ranee of Bijni was the ideal character I coveted; but I discovered that his temper was so well known that the Rajah positively declined to expose his line of elephants to an attack, which he assured me would take place if the animal became excited; in which event some valuable elephant would suffer, as the long tusks of the Bijni elephant had not been blunted, or shortened by the saw. This splendid animal was accordingly condemned to the ignominious duty of conveying food to the camp, for the other elephants upon their return from their daily work. The neighbourhood of the Brahmaputra is rich in plantain groves, and for a trifling consideration the natives allow those trees which have already produced their crop to be cut down. A full-length stem will weigh about 80 lbs., therefore an elephant is quickly loaded, as the animal for the short distance to camp will carry 18 cwts. or more. The operation of loading a pad elephant with either boughs or plantain stems is very curious. Two men are necessary; one upon the ground hands the boughs, etc., to the man upon the animal's back, who lays the thin or extreme end of the branch across the pad, leaving the thick or heavy end outwards. He places one foot upon this to keep it from slipping off until he has placed the next bough across it upon the opposite side, arranged in a similar manner. In this way he continues to load the elephant, each time holding down with his foot a separate bough, until he has secured it by the weight of another, placed in the same position opposite. This plan enables him to build up a load like a small haystack, which is then secured by ropes, and almost hides the animal that carries it. My mighty beast was condemned to this useful but degrading employment, instead of being honoured by a place in the line of shikari's elephants, and we started into the valleys among the Garo Hills, led by a native who declared that he would introduce us to rhinoceros and buffaloes.

We started at 6 A.M., and marched about 14 miles, extending into line whenever we entered a broad valley of high grass, and slowly thrashing our way through it. In many of the swampy flats among the hills the reedy grass was quite 14 or 15 feet in height and as thick as the forefinger; so dense was this herbage, that when the elephants were in line you could only see the animals upon the immediate left and right, the others being completely hidden. It struck me that this system of beating was rather absurd, as there were no stops in the front, neither scouts on the flanks, therefore any animals that might be disturbed by the advance in line had every chance of escape without being observed. The grass was a vivid green, and occasionally a rush in front showed that some large animal had moved, but nothing could be seen. This was a wrong system of beating. I was second in the line of six guns, the Rajah Suchi Khan upon my left; we presently skirted the foot of a range of low forest-covered hills, and after a rush in the high reeds I observed a couple of sambur deer, including a stag, trotting up the hill through the open forest, all of which had been recently cleared by fire. A right and left shot from Suchi Khan produced no effect, but the incident proved that the system of beating was entirely wrong, as the game when disturbed could evidently steal away and escape unseen. Our right flank had now halted at about 400 yards' distance as a pivot, upon which the line was supposed to turn in order to beat out the swamp that was surrounded upon all sides by hills and jungles. Suddenly a shot was heard about 200 yards distant, then another, succeeded by several in slow succession in the same locality. I felt sure this was a buffalo, and, as the line halted for a few minutes, I counted every shot fired until I reached the number twenty-one. Before this independent firing was completed we continued our advance, wheeling round our extreme right, and driving the entire morass, moving game, but seeing absolutely nothing. Although the jungles had been burnt, the valley grass was a bright green, as the bottom formed a swamp; even at this season (April) the ground was splashy beneath the heavy weight of our advancing line. Having drawn a blank since we heard the shots, we now assembled at the spot, where we found a bull buffalo lying dead surrounded by the elephants and four guns. These had enjoyed the fusillade of twenty-one shots before they could extinguish the old bull, who had gallantly turned to bay instead of seeking safety in retreat. It was a glorious example of the inferiority of hollow Express bullets against thick-skinned animals. The buffalo was riddled, and many of the shots were in the right place, one of which behind the shoulder would have been certain death with a solid 650 grains hard bullet, from a .577 rifle with 6 drams of powder. The buffalo, finding himself surrounded by elephants, had simply stood upon the defensive, without himself attacking, but only facing about to confront his numerous enemies.

We were a very long way from camp; we therefore retraced our course, and having avoided some dense swamps that were too soft for the elephants, we sought harder ground, shooting several hog-deer on our way, and arriving in camp after sundown, having been working for twelve hours, to very little purpose, considering our powerful equipments.

Although we had covered a very large area during the day's work, we had seen no tracks of rhinoceros, and so few of buffaloes that we determined to abandon such uninteresting and unprofitable ground; accordingly we devoted the following day to the churs or islands of the river, where we should expect no heavy game, but we might come across a tiger.

In driving the grassy islands of the Brahmaputra some persons are contented with the chance of moving tigers by simply forming a line of a quarter of a mile in length with forty elephants, without any previous arrangement or preparation. This is wrong.

To shoot these numerous islands much caution is required, and unless tigers are exceedingly plentiful, the whole day may be fruitlessly expended in marching and counter-marching under a burning sun, with a long line of elephants, to little purpose.

There should be a small herd of at least twenty head of cattle under the special charge of four shikaris, and five or six of these poor beasts should be tied up at a distance of a mile apart every evening as bait for tigers. At daylight every morning the native shikaris should visit their respective baits, and send a runner into camp with the message should one or more have been killed. The elephants being ready, no delay would occur, and the beat would take place immediately. In that manner the tiger is certain to be found, as it will be lying somewhere near the body of its prey.

There is a necessity for great precaution, lest a tiger when disturbed should steal away and escape unobserved from the dense covert of high grass. To effect his destruction, at least two scouting elephants should be thrown forward a quarter of a mile ahead from either flank of the advancing line; and, according to the conditions of the locality, two or more elephants with intelligent mahouts should be sent forward to take up positions ahead of the line at the terminus of the beat. These men should be provided with small red flags as signals should the tiger show itself; the waving of flags together with a shout will head the tiger, and drive it back towards the advancing line of elephants; at the same time the signal will be understood that a tiger is afoot, and the mahouts will be on the alert.

When a tiger is headed in this manner it will generally crouch, and endeavour to remain concealed until the elephants are close upon it. Upon such occasions it will probably spring upon the first disturber with a short harsh roar, and unless stopped or turned by a shot, it will possibly break through the line and escape to the rear, as many of the elephants will be scared and allow the enemy to pass.

Should this occur, it will be necessary to counter-march, and to reverse the position by sending some active elephants rapidly upon either flank to take up certain points of observation about 500 yards distant, according to the conditions of the ground. This forms the principal excitement of tiger-shooting in high grass, as the sport may last for hours, especially if there are only two or three guns in a long line of elephants. If there is no heavy forest at hand, but only grass jungle, no tiger should be allowed to escape if the management is good, and the patience of the hunters equal to the occasion.

I must give every credit to the Rajah Suchi Khan for this virtue, and for the perseverance he and his friends exhibited in working for so many hours in the burning sun of April to so little purpose. There was very little game upon the islands near Dhubri beyond a few hog-deer and wild pigs, and it appeared mere waste of time to wander in a long line of beating elephants from sunrise till the afternoon with scarcely a hope of tigers. However, upon the second day, when our patience was almost exhausted, we met a native who declared that a tiger had killed one of his cows only two days before. Taking him as a guide, he led us about two miles, and in a slight hollow among some green tamarisk we were, after a long search, introduced to a few scattered bones, all that remained of the native cow which had been recently killed, and the skeleton dislocated by jackals and wild pigs. Unless the tiger had been disturbed there was every chance of its being somewhere in the neighbourhood; we therefore determined to beat every yard of the island most carefully, although it extended several miles in length, and was about one mile in maximum width.

The line was formed, but no scouts were thrown forward, nor were any precautions taken; it was simply marching and counter-marching at hazard. Hours passed away and nothing was moved to break the monotony of the day but an occasional pig, whose mad rush for the moment disturbed the elephants.

It was 2 P.M.: hot work for ladies — my wife was in the howdah behind me. I confess that I am not fond of the fair sex when shooting, as I think they are out of place, but I had taken Lady Baker upon this occasion at her special request, as she hoped to see a tiger. We were passing through some dense green tamarisk, growing as close and thick as possible, in a hollow depression, which during the wet season formed a swamp, when presently the elephants began to exhibit a peculiar restlessness, cocking their ears, raising their trunks, and then emitting every kind of sound, from a shrill trumpet to the peculiar low growl like the base note of an organ, broken suddenly by the sharp stroke upon a kettle-drum, which is generally the signal of danger or alarm. This sound is produced by striking the ground with the extremity of the trunk curled up.

I felt sure that a tiger was in this dense covert. The question was how to turn him out.

The tamarisk was about 20 feet high, but the stems were only as thick as a man's arm; these grew as close together as corn in a field of wheat; the feathery foliage of green was dark through extreme density, forming an opaque mass that would have concealed a hundred tigers without any apparent chance of their discovery.

Although this depression was only about 6 feet below the general level of the island, it formed a strong contrast in being green, while the grass in the higher level was a bright yellow. The bottom had been swampy, which explained the vigorous vegetation; and although this lower level was not wider than 80 or 90 yards, it was quite a quarter of a mile in length.

Neither the mahouts nor their animals appeared to enjoy the fun of beating out this piece of dense covert, as they were well aware that the tiger was "at home." As it was absolutely necessary to form and keep a perfect line, the elephants being shoulder to shoulder, I begged the Rajah and his friends to ride towards the terminus of the tamarisk bottom, placing a gun at the extreme end and upon either side; while I should accompany the beaters to keep a correct line, and to drive the covert towards them. I felt sure that by this arrangement the tiger could not escape without being seen.

This was well carried out; they took their places, and after some delay I managed to collect about forty elephants into a straight line, not more than 4 or 6 feet from each other. The word was given for the advance, and the effect was splendid. The crash through the yielding mass was overpowering; the dark plumes of the tamarisk bowed down before the irresistible phalanx of elephants; the crackling of the broken stems was like the sound of fire rushing through a cane-brake, and this was enlivened by sudden nervous squeals, loud trumpets, sharp blows of kettle-drums, deep roars, and all the numerous sounds which elephants produce when in a state of high nervous excitement. I felt sure that at times the tiger was only a few feet in our advance, and that it was slinking away before the line.

The elephants increased in excitement; sometimes two or three twisted suddenly round, and broke the line. A halt was ordered, and although it was impossible to see beyond the animal on the immediate right and left, the order was given to dress into an exact line, and then to advance.

In this manner, with continual halts to re-form, we continued our uncertain but irresistible advance. Suddenly we emerged upon a swampy piece of grass interspersed with clumps of tamarisk; here there was intense excitement among the elephants, several turned tail and bolted in an opposite direction; when the cause was quickly discovered, by a large tiger passing exactly in front of me not 20 yards distant, and showing himself most distinctly, giving me a lovely chance.

The elephant we rode was a female named Sutchnimia, and she had been introduced to my notice as infallible, her character as usual being well supported by her mahout; but no sooner did this heroic beast descry the tiger, than she twisted herself into every possible contortion, throwing herself about in the most aimless attitudes, with a vigour that threatened the safety of the howdah and severely taxed the strength of the girth-ropes.

The tiger (a fine male) suddenly stopped, and turned three-parts round, apparently amazed at the gesticulations of the elephant; and there the beast stood, exposing the shoulder to a most certain shot if the elephant would have kept decently quiet for only two seconds. The fact of the tiger having halted, and remaining in view within 20 yards, only aggravated the terror of Sutchnimia, and she commenced shaking her colossal body like a dog that has just emerged from water. It was as much as we could do to hold on with both hands to the howdah rails; my watch was smashed, the cartridges in my belt were bent and doubled up against the pressure of the front rail and rendered useless, while the mahout was punching the head of his refractory animal with the iron spike, and the tiger was staring with astonishment at the display upon our side.

This picture of helplessness did not last long; the tiger disappeared in the dense covert, and left me to vent my stock of rage upon the panic-stricken elephant. Twice I had endeavoured to raise my rifle, and I had been thrown violently against the howdah rail, which had fortunately withstood the shock. The tiger had broken back, therefore it was necessary to repeat the beat. I was of opinion that it would be advisable to take the elephants out of the tamarisk jungle, and to march them along the open ground, so as to re-enter exactly in the same place and in the same order as before. There could be no doubt that the tiger would hold to the thick covert until fairly driven out, and it would probably break upon the second beat where the guns were protecting the end and both sides of the hollow.

The elephants were this time intensely excited, as they knew as well as we did that the game was actually before them. I ordered them to keep within a yard of each other, to make it impossible for the tiger to slink back by penetrating the line. Several times as we advanced in this close order the animal was evidently within a few feet of us, as certain elephants endeavoured to turn back, while others desired to dash forward upon the unseen danger, which all keenly smelt. At last, when several elephants trumpeted and made a sudden rush, a shot was fired from the gun upon the left flank, stationed upon the open ground slightly above the hollow. The line halted for an explanation, and it appeared that the Rajah had fired, as the tiger for an instant showed itself upon the edge of the tamarisk jungle.

We now continued the advance; the tiger had not spoken to the shot, therefore we considered that it was without effect, and I felt sure that in such compact order we should either trample upon it or push it out at the extremity of the covert.

At length, having carefully beaten out the tamarisk, which had now been almost destroyed by the tread of so close a line of elephants, we emerged at the extreme end of the hollow, where, instead of tamarisk, a dense patch of withered reeds much higher than an elephant were mingled in a confused growth, occupying an area of hardly 10 yards square. I felt sure that the tiger must have crouched for concealment in this spot.

Suchi Khan had brought his elephant upon the left, another gun was on the right, and a third in the centre at the extreme end, while I was in the bottom with the line of elephants. Begging the outside guns to be careful, and to reserve their fire until the tiger should bolt into the open, I ordered the elephants to form three parts of a circle, to touch each other shoulder to shoulder, and slowly to advance through the tangled reeds. This was well done, when suddenly the second elephant upon my left fell forward, and for the moment disappeared; the tiger had made a sudden spring, and seizing the elephant by the upper portion of the trunk, had pulled it down upon its knees. The elephant recovered itself, and was quickly brought into the position from which for a few seconds it had departed. The tiger was invisible in the dense yellow herbage.

Very slowly the line pressed forward, almost completing a circle, but just leaving an aperture a few yards in width to permit an escape. The elephant's front was streaming with blood, and the others were intensely excited, although apparently rendered somewhat confident by pressing against each other towards the concealed enemy.

Presently a mahout about two yards upon my right beckoned to me, and pointed downward with his driving-hook. I immediately backed my elephant out of the crowd, and took up a position alongside his animal. He pointed at some object which I could not distinguish in the tangled mixture of reeds, half-burnt herbage, and young green grass that had grown through; at length something moved, and I at once made out the head and shoulders of a tiger crouching as though ready for a spring. In another moment it would have tried Sutchnimia's nerves by fixing its teeth upon her trunk; but this time she stood well, being encouraged by the supporting elephants, and I placed a .577 bullet between the tiger's shoulders; this settled the morning's sport without further excitement.

The tiger was dragged out. It was a fine male, and we discovered that Suchi Khan's shot had struck it in the belly; the wound, not being fatal, had rendered it more vicious.

It has already been remarked that a really staunch and tractable elephant is rarely met with. This renders tiger-shooting exceedingly uncertain, as it is impossible to shoot correctly with a rifle when an animal is flinging itself about to an extent that renders it necessary to hold fast by the howdah rail. I generally take an ordinary No. 12 gun as an adjunct. If the grass is very high and dense, the tiger will seldom be farther than 20 yards distant, and a smooth-bore bree-chloader with a spherical ball will shoot sufficiently well to hit the palm of your hand. This accuracy may be obtained to 30 or 40 yards provided that the bullet is sufficiently large to enter the chamber, but a size too large for the muzzle. It will accordingly squeeze its way through without the slightest windage, and will shoot with great precision, with a charge of 4 1/2 drams of powder and a ball of pure soft lead. A No. 12 is exceedingly powerful, and if 7 lbs. in weight, it can be fired with one hand like a pistol. This is an immense advantage, as the shooter can hold tight by the howdah rail with his left hand, while he uses his gun with the right. I always load the right barrel with ball, and the left with the same charge of powder (4 1/2 drams), but with either 16 S.S.G. or 1 1/2 ounce of A.A. or B.B. shot. For leopards there is nothing so certain as S.S.G. at 20 or 30 yards; and for hog-deer and other sorts of small game the smaller shot is preferable, but always with the same full charge of powder.

A smooth-bore gun is much easier to use than a rifle from a howdah, as it is unnecessary to squint along the sight, but the shot is taken at once with the rapidity usual in ordinary shooting at flying objects. Care must be taken, when firing only with one hand, that the wrist should be turned to the left, so that the hammers of the gun are lying over in that direction instead of being erect. In that position the elbow is raised upon the right, and the recoil of the gun will not throw it up towards the shooter's face, which might happen should the gun be held with the hammers uppermost; it is also much easier to hold a gun with one hand in the attitude described. Should a tiger spring upon an elephant, it would be exceedingly difficult to defend the animal unless by shooting with one hand, as the struggles of the elephant would render it impossible to stand.

I had a practical example of this shortly after the departure of Suchi Khan, when I pushed on to Rohumari and met Mr. G. P. Sanderson, April 1, 1885. He had brought with him the entire force of elephants from the Garo Hills, the season for capturing wild elephants having just expired. Many of his men were suffering from fever, and he himself evidently had the poison of malaria in his system.

A bullock had been tied up the preceding evening within three-quarters of a mile from our camp, and on the morning of April 1 this was reported to have been killed. We accordingly sallied out, and in a few minutes we found the remains, above which the vultures were soaring in large numbers. The high grass had been partially burnt, and large patches remained at irregular distances where the fire had not penetrated, or where the herbage had been too green to ignite; however, all was as dry as tinder at this season, and having formed the elephants in line, I took up a position with my elephant about 300 yards ahead.

The elephants came on in excellent formation, as Mr. Sanderson was himself with them in command; presently I saw a long tail thrown up from among the yellow grass, and quickly after I distinguished a leopard moving rapidly along in my direction. For a few minutes I lost sight of it, but I felt sure it had not turned to the right or left, and, as a clump of more than ordinary thick grass stood before me, I concluded that the animal had probably sought concealment in such impervious covert.

When the elephants at length approached, I begged that half a dozen might just march through the patch within a few yards of my position. I was riding an elephant called Rosamond, which was certainly an improvement upon my former mount.

Hardly had the line entered the patch of grass when, with a short angry roar, a leopard sprang forward, and passed me at full speed within 25 yards; and immediately turned a somersault like a rabbit, with a charge of 16 S.S.G. from the No. 12 fired into its shoulder.

This was very rapidly accomplished, as our camp was within view, certainly not more than a mile distant.

We placed the leopard upon a pad elephant, and sent it home; while we once more extended the line, and as usual I took up a position some hundred yards in advance, in a spot that was tolerably clear from the high grass.

Almost the same circumstance was repeated. I saw another leopard advancing before the line, and pushing my elephant forward to a point that I considered would intercept it, I distinctly saw it enter a tangled mass of herbage, hardly large enough to shelter a calf; there it disappeared from view.

The line of elephants arrived, and no one was aware that another leopard had been moved. I pointed out the small clump of grass, and ordered an elephant to walk through it. In an instant a leopard bolted, and immediately rolled over like its comrade; but as I had to wait until it had cleared the line of elephants before I fired, it was about 35 yards distant, and although it fell to the shot, it partially recovered, and limped slowly forward with one broken leg, being terribly wounded in other places. It only went about 40 paces, and then lay down to die. One of the mahouts dismounted from his elephant, and struck it with an axe upon the head. This leopard was immediately despatched to camp, and we proceeded to beat fresh ground, as no tiger had been here, but evidently the two leopards had killed the bullock on the preceding night, and nothing more remained.

Rosamond had stood very steadily, but she was terribly rough to ride, and the howdah swung about like a boat in a choppy sea.

A couple of hours were passed in marching through every place that seemed likely to invite a tiger, but we moved nothing except a great number of wild pigs; a few of these I shot for the Garo natives who accompanied us. At length we observed in the distance the waving, green, feathery appearance of tamarisk, and as the sun was intensely hot, we considered that a tiger would assuredly select such cool shade in preference to the glaring yellow of withered grass. At all times during the hot season a dense bed of young tamarisk is a certain find for a tiger, should such an animal exist in the neighbourhood. The density of the foliage keeps the ground cool, as the sun's rays never penetrate. The tiger, being a nocturnal animal, dislikes extreme heat, therefore it invariably seeks the densest shade, and is especially fond during the hottest weather of lying upon ground that has previously been wet, and is still slightly damp; it is in such places that the tamarisk grows most luxuriantly.

We were now marching through a long strip of this character which had at one time formed a channel; on either side the tamarisk strip was enormously high, and dense grass. Suddenly an elephant sounded the kettle-drum note; this was quickly followed by several others, and a rush in the tamarisk frightened the line, as several animals had evidently broken back. We could see nothing but the waving of the bush as the creatures dashed madly past. These were no doubt large pigs, but I felt certain from the general demeanour of the elephants that some more important game was not far distant.

The advance continued slowly and steadily. Presently I saw the tamarisk's feathery tops moving gently about fifteen paces ahead of the line; the elephants again trumpeted and evinced great excitement; this continued at intervals until we at length emerged from the tamarisk upon a flat space, where the tall grass had been burnt while yet unripe, and although killed by the fire and rendered transparent, it was a mass of black and yellow that would match well with a tiger's colour. We now extended the line in more open order, to occupy the entire space of about 200 yards front; Sanderson kept his position in the centre of the line, while I took my stand in an open space about 150 yards in advance, where an animal would of necessity cross should it be driven forward by the beat.

The line advanced in good order. The elephants were much disturbed, and they evidently scented danger.

They had not marched more than 50 or 60 yards before a tremendous succession of roars scattered them for a few moments, as a large tiger charged along the line, making splendid bounds, and showing his entire length, as he made demonstrations of attack upon several elephants in quick rotation. It was a magnificent sight to see this grand animal, in the fullest strength and vigour, defy the line of advancing monsters, every one of which quailed before the energy of his attack and the threatening power of his awe-inspiring roars. The sharp cracks of two shots from Sanderson, whose elephant was thus challenged by the tiger, hardly interrupted the stirring scene; but, as the enemy rushed down the line, receiving the fire from Sanderson's howdah, he did not appear to acknowledge the affront, and having effected his purpose of paralysing the advance, he suddenly disappeared from view.

I was in hopes that he would break across the open which I commanded, but there was no sign of movement in the high grass. The line of elephants again advanced slowly and cautiously; suddenly at a signal they halted, and I observed Sanderson, whose elephant was a few yards in advance of the line, halt, and, standing up, take a deliberate aim in the grass in front. He fired; a tremendous roar was the response, and the tiger, bounding forward, appeared as though he would assuredly cross my path. Instead of this, after a rush of about 50 or 60 yards I saw the tall grass only gently moving, as the animal had reduced its pace to the usual stealthy walk. The grass ceased moving in a spot within 30 paces, and exactly opposite my position. I marked a bush upon which were a few green shoots that had sprouted since the fire had scorched the grass. I was certain that the tiger had halted exactly beneath that mark. My mahout drove the elephant slowly and carefully forward, and I was standing ready for the expected shot, keeping my eyes well open for an expected charge; Sanderson was closing in upon the same point from his position. Presently, when within a few feet of the green bush, I distinguished a portion of the tiger, but I could not determine whether it was the shoulder or the hind-quarter. Driving the elephant steadily forward, with the rifle to my shoulder, I at length obtained a complete view. The tiger was lying dead!

Sanderson's last shot had hit it exactly behind the shoulder; but the first right and left had missed when the tiger charged down the line, exemplifying the difficulty of shooting accurately with an elephant moving in high excitement.

We now loaded an elephant with this grand beast and started it off to camp, where Lady Baker had already received two leopards. We had done pretty well for the 1st April, but after this last shot our luck for the day was ended.

This day unfortunately deprived me of my companion, as the fever which had been dormant developed itself in Sanderson and completely prostrated him. He had a peculiar objection to quinine, therefore in default of remedies, which were all at hand, he remained a great sufferer during three successive weeks, and I was left alone with the long line of elephants to complete the driving of the innumerable churs below the village of Rohumari. I must pay Mr. Sanderson the well-merited compliment of praising his staff of mahouts, who were, with their well-trained animals, placed at my disposal; these men exhibited the result of such perfect discipline and organization, that, although a perfect stranger to them, I had not the slightest difficulty; on the contrary, they worked with me for twenty days as though I had been their old master for as many years. No better proof could be adduced of the excellent management of Mr. Sanderson's department.

The sport on 1st April had raised my expectations, but I quickly discovered that it was an exceptional day, and that the rule would be disappointing. A little experience introduced me to the various characters of the elephants which composed our pack, and I amused myself by arranging them according to their qualifications, the heavier and slower animals in the centre, and the more active at either end of the line. Each elephant was to retain invariably the same position every day, as the mahouts and their beasts would be more likely to act harmoniously if always associated together in the beat. The fast elephants, being at the extreme ends, would be able to turn quickly upon the centre whenever necessary. Four elephants were told off as scouts; these were the most active, with intelligent mahouts. The men appeared to take an intense interest in the sport, and in the regularity of the arrangements, as they were equally aware with myself of the necessity for strict order and discipline, where only one solitary gun represented the offensive capacity of the line.

The ordinary method of tiger-shooting with a long line of elephants comprises five or six guns placed at intervals. I dislike this style of sport, as it engenders wild and inaccurate firing. Every person wishes to secure a chance, therefore no opportunity is lost, and wherever the grass is seen to move, a bullet is directed at the spot. If only one gun is present, extreme caution and good management are necessary to ensure the death of a tiger, and the result of twenty-five days' shooting on the churs of the Brahmaputra was highly satisfactory, as during that period eight tigers and three leopards *only* were moved, and every one was bagged; thus nothing whatever escaped.

I always make a point of allowing the Government reward as a bonus, without any deductions for buffalo baits or beaters, and this amount I divide among the shikaris and mahouts according to my estimation of their merits; this gives them an additional interest in the proceedings. We were now thoroughly organised, and, if the tigers had been in the numbers that existed some years ago, we should have made a more than ordinary bag. The difficulty of managing so long a line of elephants with a tiger on foot, and only one gun, was shortly made apparent.

One of our baits had been killed, and the body had been dragged into about twelve acres of wild rose. This bush produces a blossom rather larger than the common dog-rose of English hedges, and equally lovely. Although it is armed with a certain amount of thorns, it is not to be compared with the British variety as a formidable barrier, but, as it delights in swamp hollows, it grows into the densest foliage, about 18 feet high, and forms an impenetrable screen of tangled and matted vegetation. No human being could force his way through a network of wild rose, therefore it forms a desirable retreat for all wild animals, who can penetrate beneath it, and enjoy the protection of cool shade, and undisturbed seclusion.

In an open grass country it may be readily imagined that tigers would be certain to resort to such inviting covert, where they would be secure from all intrusion, and to which cavernous density they could drag and conceal their prey.

Upon arrival about three miles from camp at this isolated patch of rose jungle, I felt sure that the tiger must be within. There was a similar but rather smaller area of wild rose about 3/4 mile distant, and it was highly probable that should the tiger be disturbed, it might slink away, break covert at the extreme end, and make off across the open grassland to the neighbouring shelter. I therefore posted myself outside the jungle in a kind of bay, where I considered the tiger would emerge from his secure hiding-place before he should risk a gallop across the open.

I threw out scouts as usual, and I sent the line of elephants round, to drive the jungle towards me from the opposite extremity.

A certain time elapsed, and at length I perceived the approach, in splendid line, each elephant as nearly as possible equidistant from its neighbour.

They marched forward in regular array until within a couple of hundred yards of my position; then suddenly I heard a trumpet, trunks were thrown up in the air, the line wavered, and a succession of well-known sounds showed that a tiger was before them. The mahouts steadied their animals, brought them again into a correct line, and the advance continued.

I was riding a large male elephant named Thompson; this was a fine animal with formidable tusks, but he was most unsteady. Already he was swaying to and fro with high excitement, as he knew full well by the trumpets and sounds of the other elephants that a tiger was not far distant.

Presently I saw the jungle shake, and a hog-deer dashed out within a few yards of me; the elephant whisked suddenly round; this prepared me for a display of his nervousness. Again the rose bushes moved, and I distinctly observed a yellowish body stealing beneath the tangled mass; it was quickly lost to sight. The line of beating elephants was coming slowly forward, crashing their way through the bush, and occasionally giving a shrill scream, when again I saw the bushes move; without further introduction a very large tigress gave two or three roars, and rushed out of the jungle exactly opposite my position, straight at my elephant. Before I had time to raise my rifle, the elephant spun round as though upon a pivot, and ran off for a few paces, making it impossible for me to fire. The tiger, probably alarmed, turned back into the secure fortress of wild rose.

We now knew that the tiger was positively between the line of elephants and myself. I felt sure that it would not show again at the same place; I therefore selected a favourable spot about 100 yards to my left upon some slightly rising ground, and the elephants wheeled and beat directly towards me.

Nothing moved except pigs, which all broke back at a wild rush between the elephants' legs, two of which had slight cuts from the tusks of boars, which had made a spiteful dig at the opposing legs whilst passing.

At length the line arrived within 20 yards from the margin of the thick jungle; here a regular rush took place; several hog-deer dashed back, but at the same time a tiger bounded forward, and galloped across the open grass-land in the direction of the neighbouring wild-rose covert. The scouts holloaed, waved their puggarees, and then rode after the tiger as hard as they could press their active elephants.

My steed Thompson had behaved disgracefully, as he had again twisted suddenly round, and was so unsteady that although the tigress was not 10 yards from me I had not the power of firing; I accordingly relinquished my favourite rifle '577, which I secured in the rack, and took in exchange my handy No. 12 smooth-bore, which only weighed 7 lbs. With that light weapon I knew I could take a quick flying shot; the right-hand barrel was loaded with a spherical ball, and the left with 1 3/4 ounce S.S.G shot and 4 1/2 drams of powder. To load a cartridge case (Kynoch's brass) with this charge, and a very thick felt wad, it is necessary to fix the wad above the shot with thick gum, otherwise it will not contain the extra quantity.

Upwards of an hour was passed in driving the second covert, but although we moved the tiger several times, it was impossible to obtain a shot, as the cunning brute, discovering our intentions, was determined not to break into the open near the elephant. At length, finding the impossibility of dislodging it, I put myself in the centre of the line, and left the end of the covert unguarded, so as to invite the tiger to make a dash through the interval to regain the former jungle.

As we marched along, driving in a compact line I presently observed the jungle move about 30 yards before me, and I immediately fired into the spot, not in the expectation of hitting an unseen animal, but I concluded that the shot would assist in driving it from the covert. This was successful, as shortly afterwards we heard the shouts of the mahouts on the scouting elephants, who reported that the tiger had gone away at great speed across the intervening ground towards the original retreat.

We hurried forward, and upon reaching the wild-rose jungle we re-formed the line, and made use of every possible manoeuvre for at least an hour without obtaining a view of the tiger. The elephants appeared confident that their enemy was there, and my men began to think that the shot I had fired into the bush might have wounded it, and that it was probably lying dead beneath some tangled foliage. By this time, through continual advancing and counter-marching, the jungle was completely trodden into confused masses of concentrated briars, which might have concealed a buffalo.

I did not share their opinion, but I concluded that the tiger was crouching, and that it would allow the elephants to pass close to its lair without the slightest movement. I accordingly ordered them to close up shoulder to shoulder, and to take narrow beats backwards and forwards to include every inch of ground. This movement was carefully worked out, and in less than fifteen minutes a sudden roar terrified the elephants, and the tiger charged desperately through the line! There was no longer any doubt about its existence, and we quickly reformed, and beat back in exactly the same close order. Twice the charge was repeated, and each time the line was broken; one elephant received a trifling scratch, and the tiger had learned that a direct charge would enable it to escape.

With only one gun it appeared to be a mere lottery, but the excitement was delightful, as there was no doubt concerning the tiger being alive, and very little doubt that it would continue its present tactics of crouching close-hidden in the dense thicket, and springing back through the line of elephants as they advanced. I now changed my position in the line, and taking with me two experienced elephants. I placed one on my right, the other on my left; we then advanced as slowly as it was possible for the elephants to move, every mahout having strict orders to keep a bright look-out, and to halt should he see the slightest movement in the bush before him. No animals were left in the jungle except the tiger, therefore any movement would be a certain sign of its presence.

We had been advancing at the rate of about half a mile an hour, the elephants almost "marking time" when in about the centre of the jungle one of the mahouts raised his arm as a signal and halted his elephant. The whole line halted immediately.

I rode towards the spot; the line opened, and the mahout explained that he distinctly saw the bushes move exactly in his front, not more than three or four paces in advance. He declared that just for one moment he had distinguished something yellow, and the tiger was in his opinion, even then, crouching exactly before us. Telling him to fall back, my two dependable elephants took their places upon the right and left. My mahout advised me not to advance, but to fire a shot into the supposed position, which he declared would either kill the tiger or drive it forward. I never like to fire at hazard, but I was of opinion that it might provoke a charge, as I did not think that anything would induce the tiger to move forward after the numerous successful attempts in breaking back. I accordingly aimed with the No.12 smooth-bore carefully in the direction pointed out by the mahout, and fired. The effect was magnificent; at the same instant a loud roar was accompanied by the determined spring of the tiger from its dense lair. My elephant twisted round so suddenly to the left, that had I been unprepared I should have fallen heavily against the rail. Instead of this, my left hand clutched instinctively the left rail of the howdah, and holding the gun with my right, I fired it into the tiger's mouth within 2 feet of the muzzle, just as it would have seized the mahout's right leg. A sack of sand could not have fallen more suddenly or heavily. The charge of S.S.G. had gone into the open jaws.

The remnant of that skull is now in my possession. The lower jaw absolutely disappeared, being reduced to pulp. All the teeth were cut away from the upper jaw, together with a portion of the bone, and several shot had gone through the back of the throat and palate into the brain. This was a striking example of the utility of a handy smooth-bore in a howdah for close quarters. If I had had my favourite '577 rifle weighing 12 lbs., I could not have used it with one hand effectively, but the 7 lb. smooth-bore was as handy as a pistol. The wind-up of the hunt was very satisfactory to my men, all of whom had worked with much intelligence and skill.

There were so many wild pigs throughout the churs below Rohumari that the tigers declined to kill our baits, as they could easily procure their much-loved food. Every night our animals were tied up in various directions, but we found them on the following morning utterly disregarded. This neglect on the part of the tigers imposed the necessity of marching in line haphazard for many hours consecutively through all the most likely places to contain a tiger. Many of the islands were at this dry season separated from each other by sandy channels where the contracted stream was only a few inches deep; it was therefore a certain proof, should tigers exist upon the islands, if tracks were discovered on the sand. During the night it was the custom of these animals to wander in all directions, and it was astonishing upon some occasions to see the great distances that the tiger had covered, and the numerous churs that it had visited, either in a search for prey, or more probably for a companion of its own species.

If there were no tracks in the channel-beds, it might be safely inferred that there were no tigers in the neighbourhood. Nevertheless I continued daily to beat every acre of ground, and we seldom returned till about 4 p.m., having invariably started shortly after daybreak.

It would be natural to suppose that the elephants would have become accustomed to the scent of tigers, from their daily occupation, and that their nerves would have been more or less hardened; but this was not the case; on the contrary, some became more restless, and evinced extreme anxiety when a pig or hog-deer suddenly rushed from almost beneath their feet. This timidity led to a serious accident, which narrowly escaped a fatal termination.

We had been fruitlessly beating immense tracts of withered grass about 10 feet high, in which were numerous pigs, but no trace of tigers, and at about noon we met some natives who were herding cattle and buffaloes. The presence of this large herd appeared to forbid the chance of finding any tigers in their vicinity, and upon questioning the herdsmen they at once declared that no such animals existed in the immediate neighbourhood; at the same time they advised us to try fresh ground upon a large island about two miles distant up the stream.

We crossed several channels, after scrambling with the usual difficulty down the cliffs, quite 35 feet high, of crumbling alluvial soil, and at length we reached the desired spot, where a quantity of tamarisk filled a slight hollow which led from the river's bed up a steep incline. By this route we ascended, and formed the elephants into line upon our left. The hollow in which my elephant remained ran parallel with the line of march, and about 5 feet below. Just as the elephants moved forward, my servant, who was behind me in the howdah, exclaimed, "Tiger, master, tiger!" and pointed to the left in the high grass a few yards in front of the line of elephants.

I could see nothing; neither could my man, but he explained that for an instant only he had caught sight of along furry tail which he was sure belonged to either a tiger or a leopard. I could always depend upon Michael, therefore I at once halted the line, with the intention of pushing my elephant ahead until I should discover some tolerably clear space among the high grass, in which I could wait for the advance of the beating line.

At about a quarter of a mile distant there was a spot where the grass had been fired while only half-ripened, and although the bottom was burnt, the stems were only scorched, and of that mingled colour, black and yellow, which matches so closely with the striped hide of a tiger. There was no better position to be found; I therefore halted, and gave the preconcerted signal for a forward movement.

The line of elephants advanced. I was riding the large tusker Thompson, who became much agitated as a succession of wild pigs rushed forward upon several occasions, and one lot took to water, swimming across a channel upon my left. Presently a slow movement disturbed the half-burnt herbage, and I could make out with difficulty some form creeping silently forward about 40 yards from my position. It halted, no doubt having perceived the elephant. It moved again, and once more halted. I now made out that it was a tiger; but although I could distinguish yellow and black stripes, I could not possibly determine any head or tail, therefore I could only speculate upon its actual attitude. It struck me that it would probably be facing me, but crouching low. The elephants were now about 150 yards distant, approaching in a crescent, as the high grass was not more than the same distance in width.

I determined to take the shot, as I felt sure that the .577 rifle would cripple the beast, and that we should find it when severely wounded; otherwise it might disappear and give us several hours' hard labour to discover. Taking a very steady aim low down in the indistinct mass, I fired.

The effect was instantaneous; a succession of wild roars was accompanied by a tremendous struggle in the high grass, and I could occasionally see the tiger rolling over and over in desperate contortions, while a cloud of black dust from the recent fire rose as from a furnace. This continued for about twelve or fifteen seconds, during which my elephant had whisked round several times and been severely punished by the driver's hook, when suddenly, from the cloud of dust, a tiger came rushing at great speed, making a most determined charge at the nervous Thompson. Away went my elephant as hard as he could go, tearing along through the grass as though a locomotive engine had left the rails, and no power would stop him until we had run at least 120 yards. During this run, with the tiger in pursuit for a certain distance, I fully expected to see it clinging to the crupper; however, by the time we turned the elephant it had retreated to the high grass covert.

I felt sure this was the wounded tiger, although Michael declared that it was a fresh animal, and that two had been together.

I now pushed the elephant into the middle of the grass, and holloaed to the line to advance in a half-circle, as I was convinced that the tiger was somewhere between me and the approaching elephants.

They came on tolerably well, although a few were rather scared. At length they halted about 70 yards from me, and, as I knew that the tiger was not far off, I ordered the left wing (on my right) to close in, so as to come round me, by which movement the tiger would be forced to within a close shot.

Before the line had time to advance, there was a sudden roar, and a tiger sprang from the grass, and seized a large muckna (tuskless male) by the trunk, pulling it down upon its knees so instantaneously that the mahout was thrown to the ground.

As quick as lightning the tiger relinquished its hold upon the elephant and seized the unfortunate mahout.

I never witnessed such a hopeless panic. The whole line of elephants broke up in complete disorder. The large elephant Hogg, who had been seized, was scaring riderless at mad speed over the plain; a number of others had bolted in all directions, and during this time a continual succession of horrible roars and angry growls told that the tiger was tearing the man to pieces. A cloud of dust marked the spot within 70 paces of my position. It was like a dreadful nightmare; my elephant seemed turned to stone. In vain I seized the mahout by the back of the neck and nearly dislocated his spine in the endeavour to compel him to move forward; he dug his pointed hook frantically into Thompson's head, but the animal was as rigid as a block of granite. This lasted quite fifteen seconds; it appeared as many minutes. Suddenly my servant shouted "Look out, master, another tiger come; two tigers, master, not one!" I looked in the direction pointed, and I at once saw a tiger crouching as though preparing for a charge, about 40 yards distant: the animal was upon my right, and the elephant had not observed it.

I fired exactly below the nose, and the tiger simply rolled upon its side stone-dead, the bullet having completely raked it. Leaving the body where it lay, my elephant now responded to the driver's hook, and advanced steadily towards the spot where we had seen the cloud of dust which denoted the attack upon the mahout. Fully expecting to see the tiger upon the man's body, I was standing ready in the howdah prepared for a careful shot. We arrived at the place. This was cleared of grass by the recent struggle, but instead of finding the man's body, we merely discovered his waist-cloth lying upon the ground a few yards distant. About 15 yards from this bloody witness we saw the unfortunate mahout lying apparently lifeless in the grass.

We immediately carried him to the river and bathed him in cool water. He had been seized by the shoulder, and was terribly torn and clawed about the head and neck, but fortunately there were no deep wounds about the cavity of the chest. We bandaged him up by tearing a turban into long strips, and having made a good surgical job, I had him laid upon a pad elephant and sent straight into camp. We then loaded an elephant with the tiger, which we proved to be the same and only animal (a tigress) which had charged the elephant after my first shot. The bullet had struck the thigh bone, causing a compound fracture, and that accounted for the escape of Thompson without being boarded from the rear, as she could not spring so great a height upon only three legs. The furious beast had then attacked the elephant named Hogg, which, falling upon its knees, had thrown the unready driver. We subsequently discovered that he had a boil upon his right foot, which had prevented him from using the rope stirrup; this accounted for the fall from his usually secure seat.

The tigress, having mauled her victim and left him for dead, was prepared for an onset upon Thompson had I not settled her with the .577 bullet in the chest.

On arrival at the camp the man was well cared for, and on the following morning we forwarded him by boat to the hospital at Dhubri in charge of the keddah doctor. It was satisfactory to learn that after a few months he recovered from his wounds, and exhibited his complete cure by absconding from the hospital unknown to the authorities, without returning thanks for the attention he had received.

This incident was an unfortunate example of the panic that can be established among elephants. It is a common saying that the elephant depends upon the mahout; this is the rule for ordinary work, but although a staunch elephant might exhibit nervousness with a timid mahout, no driver, however determined, can induce a timid animal to face a tiger, or to stand its onset. Thompson had behaved so badly that I determined to give him one more chance, and to change him for another elephant should he repeat his nervousness.

A few days after this occurrence, the natives reported a tiger to be in a thicket of wild rose. We had changed camp to a place called Kikripani, about eight miles from Rohumari, and I immediately took the elephants to the wild-rose jungle, which was about two miles distant.

The usual arrangements were made, and I took up a position upon Thompson in a narrow opening of fine grass which cut at right angles through the wild-rose thicket. As the elephants approached in close order, I was certain, from the peculiar sounds emitted, that a tiger or some unbeloved animal was before them, and upon the advance of the line to within 30 yards of the open ground a rustling in the bush announced the presence of some animal which could not much longer remain concealed. Suddenly a large panther bounded across the open, and I took a snap shot, which struck it through the body a few inches behind the shoulder. It rolled over to the shot, but immediately disappeared in the thick jungle a few paces opposite.

I called the line of elephants, and we lost no time in beating the neighbouring bush in the closest order, as I fully expected the panther would be crouching beneath the tangled mass of foliage.

In a short time the elephants sounded, and without more ado the panther forsook its cover and dashed straight at Thompson, seizing this large elephant by the shoulder joint, and hanging on like a bulldog with teeth and claws. Away went Thompson through the tangled rose-bushes, tearing along like a locomotive! It was impossible to fire, as the panther was concealed beneath the projecting pad below the howdah, and I could not see it. In this manner we travelled at railway pace for about 100 yards, when I imagine the friction of the thick bush through which we rushed must have been too much for the resistance of the attacking party, and the panther lost its hold; in another instant it disappeared in the dense jungle.

I now changed my elephant, and rode a steady female (Nielmonne), and the line having re-formed, we advanced slowly through the bush. We had not gone 50 yards before the elephants scented the panther, and knowing the stealthy habits of the animal I formed a complete circle around the spot, and closed in until we at length espied the spotted hide beneath the bush. A charge of buckshot killed it without a struggle.

According to my own experience, there can be no comparison in the sport of hunting up a tiger upon a good elephant in open country, and the more general plan of driving forest with guns placed in position before a line of beaters. By the former method the hunter is always in action, and in the constant hope of meeting with his game, while the latter method requires much patience, and too frequently results in disappointment. Nevertheless, to kill tigers, every method must be adopted according to the conditions of different localities.

Under all circumstances, if possible, a dependable elephant should be present, as many unforeseen cases may arrive when the hunter would be helpless in the absence of such an animal; but, as we have already seen, the danger is extreme should the elephant be untrustworthy, as a runaway beast may be an amusement upon open grass-land, but fatal to the rider in thick forest.

The only really dependable elephant that I have ever ridden was a tusker belonging to the Commissariat at Jubbulpur in 1880; this fine male was named Moolah Bux. He was rather savage, but he became my great friend through the intervention of sugar-canes and the sweet medium of jaggery (native sugar) and chupatties, with which I fed him personally whenever he was brought before me for the day's work; I also gave him some bonne-bouche upon dismounting at the return to camp.

Although Moolah Bux was the best elephant I have myself experienced, he was not absolutely perfect, as he would not remain without any movement when a tiger charged directly face to face; upon such occasions he would stand manfully to meet the enemy, but he would swing his huge head in a pugnacious spirit preparatory to receiving the tiger upon his tusks.

The first time that I witnessed the high character of this elephant was connected with a regrettable incident which caused the death of one man and the mutilation of two others, who would probably have been killed had not Moolah Bux been present. The description of this day's experience will explain the necessity of a staunch shikar elephant when tiger-shooting, as the position may be one that would render it impossible to approach on foot when a wounded and furious tiger is in dense jungle, perhaps with some unfortunate beater in its clutches.

I was shooting in the Central Provinces, accompanied by my lamented friend the late Mr. Berry, who was at that time Assistant-Commissioner at Jubbulpur.

We were shooting in the neighbourhood of Moorwarra, keeping a line as nearly as possible parallel with the railway, limiting our distance to 20 miles in order to obtain supplies. This arrangement enabled us to receive 30 lbs. of ice daily from Allahabad, as a coolie was despatched from the station immediately upon arrival of the train, the address of our camp being daily communicated to the stationmaster. It was the hot season in the end of April, when a good supply of ice is beyond price; the soda-water was supplied from Jubbulpur, and with good tents, kuskos tatties, and cool drinks, the heat was bearable. It was this heat that had brought the tigers within range, as all water-springs and brooks were dried up, the tanks had evaporated, and the only water procurable was limited to the deep holes in the bends of streams that were of considerable importance in the cooler seasons of the year. The native headmen had received orders from the Deputy-Commissioner to send immediate information should any tigers be reported in their respective districts; they had also received special instructions to tie up buffaloes for bait should the tracks of tigers be discovered. The latter order was a mistake, as the buffaloes should not have been tied up until our arrival at the locality; upon several occasions the animals were killed and eaten some days before we were able to arrive upon the scene.

This was proved to be the case upon our arrival at Bijore, about nine miles from the town of Moorwarra, where the zealous official had exhibited too eager a spirit for our sport. Two buffaloes had been tied up about half a mile apart, near the dry bed of a river, where in an abrupt bend the current had scooped out a deep hole in which a little water still remained. Both buffaloes had been killed, and upon our arrival early in the morning nothing could be discovered except a few scattered bones and the parched and withered portions of tough hide.

There were tracks of tigers upon the sand near the drinking-place, also marks of cheetul and wild pigs, therefore we determined to drive the neighbouring jungle without delay.

The neighbourhood was lovely, a succession of jungles and open grass-glades, all of which had been burnt clean, and exceedingly fine grass, beautifully green, was just appearing upon the dark brown surface scorched by the recent fire.

There were great numbers of the ornamental mhowa trees, which from their massive growth resembled somewhat the horse-chestnut trees of England. These had dropped their luscious wax-like blossoms, which from their intense sweetness form a strong attraction to bears and other animals of the forests; they also form a valuable harvest for the natives, who not only eat them, but by fermentation and distillation they produce a potent spirit, which is the favourite intoxicating liquor of the country.

If game had been plentiful this would have been a charming hunting-ground, but, like most portions of the Central Provinces, the animals have been thinned by native pot-hunters to an extent that will entail extermination, unless the game shall be specially protected by the Government. When the dry season is far advanced, the animal can only procure drinking water at certain pools in obscure places among the hills; these are well known to the native sportsman, although concealed from the European. On moonlight nights a patient watch is kept by the vigilant Indian hunter, who squats upon a mucharn among the boughs within 10 yards of the water-hole, and from this point of vantage he shoots every animal in succession, as the thirst—driven beasts are forced to the fatal post.

Nothing is more disappointing than a country which is in appearance an attractive locality for wild animals, but in reality devoid of game. I make a point of declining all belief in the statements of natives until I have thoroughly examined the ground, and made a special search for tracks in the dry beds of streams and around the drinking-places. Even should footprints be discovered in such spots, they must be carefully investigated, as the same animals visit the water-hole nightly, and in the absence of rain, the tracks remain, and become numerous from repetition; thus an inexperienced person may be deceived into the belief that game is plentiful, when, in fact, the country contains merely a few individuals of a species. It must also be remembered that during the dry season both deer, nilgyhe, and many other animals travel long distances in search of water, and return before daylight to their secluded places of retreat.

This was the position of Bijore at the period of our visit; the most lovely jungles contained very little game. Although our baits had been devoured some days ago, I could not help thinking that the tiger might still be lurking in the locality, as it had been undisturbed, and there was little or no water in the neighbourhood excepting one or two drinking places in the beds of nullahs.

We had 164 beaters, therefore we could command an extensive line, as the jungles, having been recently burnt, were perfectly open, and an animal could have been seen at a distance of 100 yards.

Having made all the necessary arrangements, the beat commenced. It was extraordinary that such attractive ground contained so little game. The surface was a delicate green from the young shoots of new grass, and notwithstanding the enticing food there were no creatures to consume the pasturage.

Hours passed away in intense heat and disappointment; the most likely jungles were beaten with extreme care, but nothing was disturbed beyond an occasional peacock or a scared hare. The heat was intense, and the people having worked from 6 a.m. began to exhibit signs of weariness, as nothing is so tiring as bad luck. Although the country was extremely pretty it was very monotonous, as each jungle was similar in appearance, and I had no idea how far we were from camp; to my surprise, I was informed that we had been working almost in a circle, and that our tents were not more than a mile and a half distant in a direct line. We came to the conclusion that we should beat our way towards home, carefully driving every jungle in that direction.

During the last drive I had distinctly heard the bark of a sambur deer about half a mile in my rear, which would be between me and the direction we were about to take. It is seldom that a sambur barks in broad daylight unless disturbed by either a tiger or leopard; I was accordingly in hope that the sound might be the signal of alarm, and that we might find the tiger between us and the neighbouring village by our camp, where a small stream might have tempted it to drink.

Having taken our positions-Mr. Berry amidst a few trees which formed a clump in a narrow glade outside, and myself around the corner of a jungle — the beat commenced. I was in the howdah upon Moolah Bux, and from my elevated position I could look across the sharp corner of the jungle and see a portion of the narrow glade commanded by my companion Berry; upon my side there was a large open space perfectly clear for about 200 yards, therefore the jungle was well guarded upon two sides, as the drive would terminate at the corner.

In a short time the usual monotony of the beater's cries was exchanged for a series of exciting shouts, which showed that game of some kind was on foot. We had lost so much hope, that the presence of a tiger was considered too remote to restrict our shooting to such noble game, and it had been agreed to lose no chance, but to fire at any animal that should afford a shot. Presently, after a sudden roar of animated voices, I saw ten or twelve wild pigs emerge from the jungle and trot across the glade which Berry commanded. A double shot from his rifle instantly responded.

The line of beaters was closing up. This was a curious contrast to the dull routine which had been the character of the drives throughout the day; there was game afoot, and the jungle being open, it could be seen, therefore immense enthusiasm was exhibited by the natives. Another burst of excited voices proclaimed a discovery of other animals, and a herd of eight or ten spotted deer (cheetul) broke covert close to my elephant and dashed full speed across the open glade. They were all does and young bucks without antlers, therefore I reserved my fire. We could not now complain of want of sport, as all the animals appeared to be concentrated in this jungle; another sudden yelling of the beaters was quickly followed by a rush of at least twenty pigs across Berry's glade, and once again his rifle spoke with both barrels in quick succession. I was in hope that the sambur stag that I had heard bark in this direction might be still within the drive, but the beaters were closing up, and the greater portion of the line had already emerged upon either side of the acute angle.

I now perceived Berry advancing towards me, he having left his place of concealment in the clump of trees. "Did you see him?" he exclaimed, as he approached within hearing distance. "See what?" I replied; "have you wounded a boar?" "A boar! No; I did not fire at a boar, but at a tiger, the biggest that I ever saw in my experience! He passed close by me, within 20 yards, at the same time that the herd of pigs broke covert; and I fired right and left, and missed him with both barrels; confound it."

This was a most important announcement, and I immediately dismounted from my elephant to examine the spot where the tiger had so recently appeared. It must indeed have been very close to Berry, as I had not seen the beast, my line of view being limited by the intervening jungle to the portion of the glade across which the pigs had rushed.

I now measured the distance from Berry's position to the tracks of the tiger, which we discovered after some few minutes' search. This was under 20 yards. The question now most important remained—Was the tiger wounded? A minute investigation of the ground showed the mark of a bullet, but we could find no other. This looked as though it must have struck the tiger, but Berry was very confident that such was not the case, as he declared the tiger did not alter his pace when fired at, but, on the contrary, he walked majestically across the narrow glade with his head turned in the opposite direction from Berry's position. He was of opinion that the tiger had not been disturbed by the close report of the rifle, as the noise of 164 beaters shouting at the maximum power of their voices was so great that the extra sound of the rifle bore only a small proportion.

We looked in vain for blood-tracks, and having come to the conclusion that Berry had fired too high in a moment of excitement, we now made the most careful arrangements for driving the jungle into which the tiger had so recently retreated.

This formed a contrast to all others that we had beaten during the morning's work, as it had not been burnt. The fire had stopped at a native footpath, and instead of the bare ground, absolutely devoid of grass or dead leaves, the withered herbage as yellow as bright straw stood 3 feet high, and formed a splendid cover for animals of all kinds. I felt certain that the tiger would not leave so dense a covert without an absolute necessity; at the same time it was necessary to make a reconnaissance of the jungle before we could determine upon our operations.

Mounting my elephant Moolah Bux, I begged Berry to take Demoiselle, and accompanied by a couple of good men we left the long line of beaters stationed in order of advance along the glade, with instructions to march directly that we should send them the necessary orders. I begged them upon this occasion not to shout, but merely to tap the trees with their sticks as their line came forward.

We proceeded about a quarter of a mile ahead, and then turned into the jungle on our left. Continuing for at least 300 yards, we arrived at some open ground much broken by shallow nullahs, which formed natural drains in a slight depression of grassy land between very low hills of jungle, through which we had recently passed. There was a small nullah issuing from the forest, in which I placed my elephant, and I begged my friend Berry to ride Demoiselle to a similar place about 200 yards upon my right. I concluded that should the tiger be between us and the line of beaters, he would in all probability steal along one or the other of these nullahs before he could cross the open ground. We now sent back one of the natives with orders for the line of beaters to advance. Mr. Berry left upon

Demoiselle to take up his position, while I pushed Moolah Bux well into the jungle in the centre of the small nullah, which commanded a clear view of about 20 yards around.

In a short time we heard the clacking sound of many sticks, the beaters having obeyed the injunction, and keeping profound silence with their voices.

There were no animals in this jungle, probably they had been frightened by the great noise of the beaters when shouting in the recent drive; at any rate, the beat was barren, and having waited fruitlessly until I could see the men approaching within a few yards of my position, I ordered the elephant to turn round, with the intention of proceeding another quarter of a mile in advance, and thus continuing to beat the jungle in sections until it should be thoroughly driven out.

I had hardly turned the elephant, when we were startled by tremendous roars of a tiger, continued in quick succession within 50 yards of the position that I occupied. I never heard either before or since such a volume of sound proceeding from a single animal; there was a horrible significance in the grating and angry voice that betokened the extreme fury of attack. Not an instant was lost! The mahout was an excellent man, as cool as a cucumber, and never over-excited. He obeyed the order to advance straight towards the spot, in which the angry roars still continued without intermission.

Moolah Bux was a thoroughly dependable elephant, but although moving forward with a majestic and determined step, it was in vain that I endeavoured to hurry the mahout; both man and beast appeared to understand their business thoroughly, but to my ideas the pace was woefully slow if assistance was required in danger.

The ground was slightly rising, and the jungle thick with saplings about 20 feet in height, and as thick as a man's leg; these formed an undergrowth among the larger forest trees.

Moolah Bux crashed with ponderous weight through the resisting mass, bearing down all obstacles before him as he steadily made his way through the intervening growth. The roars had now ceased. There were no leaves upon the trees at this advanced season, and one could see the natives among the branches in all directions as they were perched for safety in the tree-tops, to which they had climbed like monkeys at the terrible sounds of danger. "Where is the tiger?" we shouted to the first man we could distinguish in this safe retreat only a few yards distant. "Here, here!" replied the man, pointing immediately beneath him. Almost at the same instant, with a loud roar, the tiger, which had been lying ready for attack, sprang forward directly for Moolah Bux.

There were so many trees intervening that I could not fire, and the elephant, instead of halting, moved forward, meeting the tiger in its spring. With a swing of his huge head Moolah Bux broke down several tall saplings, which crashed towards the infuriated tiger and checked the onset; whether the animal was touched by the elephant's tusks I could not determine, but it appeared to be within striking distance when the trees were broken across its path. Discomfited for the moment, the tiger bounded in retreat, and Moolah Bux stood suddenly like a rock, without the slightest movement. This gave me a splendid opportunity, and the '577 bullet rolled the enemy over like a rabbit. Almost at the same instant, having performed a somersault, the tiger disappeared, and fell struggling among the high grass and bushes about 15 paces distant.

I now urged Moolah Bux carefully forward until I could plainly see the tiger's shoulders, and a second shot through the exact centre of the blade-bone terminated its existence.

The elephant had behaved beautifully, and I have frequently looked back to that attack in thick forest, and been thankful that I was not mounted upon such animals as I have since that time had the misfortune to possess. Moolah Bux now approached the dead body, and at the command of the mahout he pulled out by the roots all the small undergrowth of saplings and dried herbage to clear a space around his late antagonist. In doing this his trunk several times touched the skin of the tiger, which he appeared to regard with supreme indifference.

I gave two loud whistles with my fingers as a signal that all was over, and we were still occupied in clearing away the smaller growth of jungle, when a native approached as though very drunk, reeling to and fro, and at length falling to the ground close to the elephant's heels; the man was covered with blood, and he had evidently fainted. I had an excellent Madras servant named Thomas, who was behind me in the howdah, and he lost no time in descending from the elephant and in pouring water over the unfortunate coolie, from a jar which I handed from beneath the seat. In a few moments the man showed signs of life, and the beaters began to collect around the spot. Two men were approaching supporting a limp and half-collapsed figure between them, completely deluged with blood; this was a second victim of the tiger's attack. Both men were now laid upon the ground, and water poured over their faces and chests; but during this humane operation another party was observed, carrying in their arms the body of a third person, which was hardly to be recognised through the mass of blood coagulated and mixed with dead leaves and sand, as the tiger had dragged and torn its victim along the ground with remorseless fury. This was a sad calamity. There could be little doubt that when we heard the roars of the infuriated beast it was attacking the line of beaters, and knocking them over right and left before they had time to ascend the trees. The village was only a mile distant, and we immediately sent for three charpoys (native bed-steads) as stretchers to convey the wounded men. Demoiselle arrived with Mr. Berry, who came into my howdah, while the tiger was with some difficulty secured upon the pad of that exceedingly docile elephant. In this form we entered the village as a melancholy procession;, the news having spread, all the women turned out to meet us, weeping and wailing in loud distress, and the scene was so touching that I began to reflect that tiger-shooting might be fun to some, but death to others, who, poor fellows, had to advance unarmed through dangerous jungle.

The reason for this savage attack was soon discovered. As a rule, there is little danger to a line of beaters provided the tiger is unwounded, and no person should ever place his men in the position to drive a jungle when a wounded tiger is in retreat. In such a case, if no elephants are present, it would be necessary to obtain the assistance of buffaloes; a herd of these animals driven through the jungle would quickly dislodge a tiger. We now skinned our late enemy, while a messenger was started towards Moorwarra, 9 or 10 miles distant, to prepare the authorities for the reception of our wounded men in hospital.

The skin having been taken off, we discovered a small hole close to the root of the tail, which had not been observed. Upon a close examination with the finger, I found minute fragments of lead, resembling very small shot flattened upon an anvil. The hole was not deeper than 1 1/4 inch in the hard muscle of the rump, and the only effect of Berry's '577 hollow Express was to produce this trumpery wound, which had enraged the animal without creating any serious injury. It is necessary to explain that the bullet of this rifle was more than usually light and hollow; but the want of penetrating power of the hollow projectile, and the dangerous results, were terribly demonstrated, notwithstanding the large charge of 6 drams of powder.

A comparison of the effect of my '577 with the same charge of 6 drams, but with a solid bullet of ordinary pure lead weighing 648 grains, was very instructive. The first shot, when the tiger was bounding in retreat after it had charged the elephant, had struck the right flank, and as the animal was moving obliquely, the bullet had passed through the lungs, then, breaking the shoulder-bone, it was found in its integrity just beneath the skin of the shoulder upon the side opposite to that of entry; it was very much flattened upon one side, as it had traversed an oblique course throughout, and had torn the inside of the animal in a dreadful manner. The second shot, fired simply to extinguish the dying tiger, passed through both shoulders, but was found under the skin upon the opposite side, flattened exactly like a mushroom, into a diameter of about 1 1/2 inch at the head, leaving about half an inch of the base uninjured which represented the stalk. This was a large tiger, and remarkably thick and heavy, with strong and hard muscles, nevertheless the penetration of the soft leaden bullet was precisely correct for that quality of game. If the '577 bullet had been made of an admixture of tin or other alloy to produce extreme hardness, it would have passed through the body of the tiger with a high velocity, but the animal would have escaped the striking energy, which would not have been expended upon the resisting surface. It is the striking energy, the knocking-down power of a projectile, that is so necessary when hunting dangerous game. I cannot help repetition in enforcing this principle: there is a minimum amount of striking energy in a light hollow projectile, and a maximum amount in a solid heavy projectile; keep the latter within the animal to ensure the effect of the blow; this will be effected by a bullet made of pure lead without admixture with other metal, to flatten upon impact, and by the expansion of surface it will create a terrific wound; at the same time it will have sufficient momentum from its great weight to push forward, and to overcome the resistance of opposing bones and muscles. A very large tiger may weigh 450 lbs.; a '577 bullet of 650 grains, propelled by 6 drams of powder, has a striking energy of 3520 foot-pounds. This may be only theoretical measurement, but the approximate superiority of 3500 lbs. against the tiger's weight, 450 lbs., would be sufficient to ensure the stoppage of a charge, or the collapse of the animal in any position, provided that the bullet should be retained within the body, and thus bestow the whole force of the striking energy.

We did all that could be done for our wounded men. The strength of caste prejudices was so potent that, although in pangs of thirst from pain and general shock to the system, they would accept nothing from our hands. I made a mixture of milk with soda-water, brandy, and laudanum, but they refused to swallow it, and the only course, after washing their wounds and bandaging, was to leave them to the treatment of their own people.

One man was severely bitten through the chest and back, the fangs of the tiger having penetrated the lungs; he was also clawed in a terrible manner about the head and face, where the paws of the animal had first made fast their hold. This man died in a few hours. The others were bitten through the shoulder and upper portion of the arm, both in the same manner, and the sharp claws had cut through the scalp from the forehead across the head to the back of the neck, inflicting clean wounds to the bone, as though produced by a pruning-knife. They were conveyed in litters to the hospital in Moorwarra, where they remained for nearly a month, at the expiration of which they recovered. The seizure by the claws was effected without the shock of a blow.

This serious accident was entirely due to a hollow bullet: if a solid bullet had struck a tiger in the same place it would have carried away a portion of the spine, and the animal would have been paralysed upon the spot.

In the absence of a dependable elephant we should have been helpless, and the tiger might have wounded or killed many others.

CHAPTER VII

THE TIGER (CONTINUED)

The day after the accident described, we were sitting beneath the shade of a mango grove at about 4 P.M. when a native arrived at the camp with news that a tiger had just killed a valuable cow which gave him a large supply of milk, and the body was lying about 2 miles distant. The tragic incident of the previous day had establis-hed a panic in the village, and the natives were not in the humour to turn out as beaters. I quite shared their feeling, as I did not wish to expose the poor people after the loss they had sustained; it was too late for a beat, therefore I determined to take the two elephants and make a simple reconnaissance, that might be of use upon the follo-wing day.

It was 4.30 P.M. by the time we started, as the two elephants had taken some time to prepare. The native was tolerably correct in his estimate of distance, and after passing through a long succession of glades and wooded hills, broken by deep nullahs, we arrived at the place, where soaring vultures marked the spot, and the remains of a fine white cow were discovered, that had been killed upon the open ground and dragged into the dense jungle. Leaving Demoiselle in the open, and taking Berry into my howdah upon Moolah Bux, we carefully searched the jungle until sunset, but finding nothing, we were obliged to return to camp, having made ourselves thoroughly acquainted with the conditions of the locality. On the following morning at daylight I took only twenty men, who had recovered from their panic, and with the two elephants and a very plucky policeman we made our way to the place where the body of the cow was lying on the previous evening. It was gone. Leaving all the men outside the jungle, we followed on Moolah Bux, tracking along the course where the tiger had dragged the carcase, and keeping a sharp look-out in all directions. After a course of about 150 yards we arrived at a spot where the tiger had evidently rested: here it had devoured the larger portion, and nothing but the head remai-ned. It was impossible to decide whether jackals or hyenas had made away with the remnants, or whether the tiger had carried them off to some secure hiding-place, but it was highly probable that the animal was not far distant.

The jungle was not more than 5 or 6 acres, and it was surrounded by grass; we therefore determined to arrange scouts around, while we should thoroughly but slowly examine the covert upon the two elephants.

There was nothing in the drive.

The slope upon which the jungle was situated drained towards an exceedingly deep and broad nullah; this formed the main channel, into which numerous smaller nullahs converged from the surrounding inclination. The general character of the country was withered grass upon numerous slopes, the tops of which were covered with low jungle. At the lower portion of the deep nullah there was a small but important pool of water, as it was the only drinking-place within a distance of 2 miles. As usual, there was a sandbank around this deep pool, which, being in the bend of the nullah, had been swept out of the opposing bank and deposited near, the drinking-hole. Upon this sandy surface we found several tracks of tigers, and we arrived at the conclusion that a tiger and tigress had been together, and that I had killed the male on the occasion of the accident; the female would therefore be the animal of which we were in search.

The nullah was about 20 yards across and 30 feet in depth; the banks were in most places perpendicular, and the bottom was rough with stones, intermingled with bushes, most of which had lost their foliage. It was quite possible that, after drinking the tigress might have lain down to sleep among the bushes, where the hollowed bank afforded a cool shade; but I did not like to send men into the dangerous bottom, and the banks were so steep that the elephants could not possibly descend.

About 400 paces distant, a large tree grew from the right bank, and the branches overhung the nullah; I therefore suggested to Berry that he should take up a position in the boughs, and that we would beat towards him by pelting the bottom of the ravine with stones; should the tigress break back, I could stop her from the howdah, and should she move forward, she must pass directly beneath the tree upon which Berry would be seated. This plan was carried out, but the plucky policeman insisted upon descending into the nullah and walking up the bottom, while the natives upon either side bombarded the banks with stones.

There was absolutely nothing alive in that inviting nullah. I had walked Moolah Bux slowly along, looking down from the margin of the ravine, and upon arrival at Berry's perch I took him up behind me in the rear compartment of the howdah. I felt almost sure that, although we had drawn a blank up to the present time, the tigress would be lying somewhere among the numerous deep but narrow nullahs which drained into the main channel that we had just examined. We therefore determined to leave all the men seated upon a knoll on the highest ground, while we should try the various nullahs upon Moolah Bux; as he could walk slowly along the margin so close to the edge that we should be able to look down into the bottom of each ravine, and in the parched state of vegetation nothing could escape our view.

The natives were well satisfied with this arrangement, and they took their seats upon a grassy hill, which afforded a position from which they could watch our movements.

Moolah Bux commenced his stately march, walking so close to the hard edge of the deep nullahs that I was rather anxious lest the bank should suddenly give way. The instinct of an elephant is extraordinary in the selection of firm ground. Although it appeared dangerous to me, Moolah Bux was perfectly satisfied that the ground would bear his weight, and he continued his risky march, both up and down a number of those monotonous ravines which scored the slopes in all directions, but without success.

The sun was like fire, and it was difficult to grasp the barrel of the rifle. It was past noon, and we had been working unceasingly since 6 a.m. The bottoms of the ravines were filled some feet in depth with dry leaves, which had fallen from the trees (now naked) which fringed the banks, therefore we could have seen a cat had she been lying either in the nullah or upon the barren sides. "There is no tigress here," said Berry; "this is one of those sly brutes, that kills and eats, but does not remain near her kill; she is probably a couple of miles away while we are looking for her in these coverless nullahs."

These words were hardly uttered, when we suddenly heard a rushing sound like a strong wind, which seemed to disturb the dried leaves in the deep bottom somewhere in our front. At first I could hardly understand the cause, but in a few seconds a large tigress sprang up the bank, and appeared about 20 paces in our front. Without a moment's hesitation she uttered several short roars, and upon the beautifully clean ground she bounded forward in full charge straight for Moolah Bux. I never saw a more grand but un-provoked attack.

The elephant was startled by the unexpected apparition, and I could not fire, as he swung his mighty head upon one side, but almost immediately he received the tigress upon his long tusks, and with a swing to the right he sent her flying into the deep nullah from which she had just emerged.

Although the trees and shrubs were utterly devoid of leaves, the-re was unfortunately a large and dense evergreen bush exactly op-posite, called karoonda; the tigress sprang up the bank, and disap-peared behind this opaque screen before we had time to fire.

The mahout, who was a splendid fellow, perceived this in an in-stant, and driving his elephant a few paces forward, he turned his head to the right, giving me a beautiful clear sight of the tigress, bounding at full speed about 80 paces distant along the clean surface of parched herbage, up a slight incline.

I heard the crack of Berry's rifle close to my ear, but no effect was produced. The tigress was going directly away from us, and Moolah Bux stood as firm as a rock, without the least vibration. As I touched the trigger, the tigress performed a most perfect somersault, and lay extended on the bare soil with her head turned towards us, and her tail stretched in a straight line exactly in the opposite direction. A great cheer from our men, who had witnessed the flying shot from their position on the knoll, was highly satisfactory.

We now turned back, and at length discovered a spot where the elephant could descend and cross the deep nullah. We then mea-sured the distance—82 yards, as nearly as we could step it. My .577 solid bullet of pure lead had struck the tigress in the back of the neck; it had reduced to pulp several of the vertebrae, and entering the brain, it had divided itself into two portions by cutting its substance upon the hard bones of the broken skull, which was literally smashed to pieces.

I found a sharp-pointed jagged piece of lead, representing about one-third of the bullet, protruding through the right eye-ball; the remaining two-thirds I discovered in the bones of the face by the back teeth, where it was fixed in a misshapen but compact mass among splinters of broken jaw.

Berry's bullet had also struck the tigress, but precisely in the same place, close to the root of the tail, where he had wounded the tiger a short time before. Upon arrival at the camp we skinned the animal, and took special pains to prove the effect of the unfortunate hollow bullet. This was conclusive, and a serious warning.

The penetration was only an inch in depth. We washed the flesh in cold water, and searched most carefully throughout the lacerated wound, which occupied a very small area of about 1 inch. In this we found two pieces of the copper plug which stopped the hole in front of the bullet, together with a number of very minute fragments or flakes of lead; these proved that the extremely hollow projectile had broken up, and was rendered abortive almost immediately upon impact.

The danger of such a bullet was manifest; it was almost as hollow as a hat, and almost as harmless as a hat would be, if thrown at a charging tiger.

This was an interesting exception to the rule that is generally accepted, that a tiger will not attack if left undisturbed. If any person had been walking along the margin of that nullah, he would have been seized and destroyed without doubt by that ferocious beast. There was a case in point last year (1888) in the Reipore district, when Mr. Lawes, the son of the missionary of that name, was killed by a tigress, which was the first to attack. This animal was reported by the natives to be in a certain nullah within a short distance of the camp. The young man, who was quite inexperienced, took a gun, and with a few natives proceeded to the spot on foot. Looking over the edge of the nullah in the hope of finding the tiger lying down, he was suddenly startled by an unexpected attack; a tigress bounded up the steep bank and seized Mr. Lawes before he had time to fire. The animal did not continue the attack, but merely shook him for a few moments, and then retreated to her lair; he was so grievously wounded that he died on the following day, after his arrival in a litter at Reipore.

Many people imagine that a tiger attacks man with the intention of eating him, as a natural prey; this is a great mistake. The greater number of accidents are occasioned by tigers which have no idea of making a meal of their victims; they may attack from various reasons. Self-defence is probably their natural instinct; the tiger may imagine that the person intends some injury, and it springs to the attack; or it may be lying half asleep, and when suddenly disturbed it flies at the intruder without any particular intention of destroying him, but merely as a natural result of being startled from its rest. When, driven by a line of beaters, the tiger breaks back, it may be readily understood that it will attack the first individual that obstructs its retreat, but in no case will the tiger eat the man, unless it is a professional man-eater.

The cunning combined with audacity of some man-eaters is extraordinary.

A few years ago there was a well-known tiger in the Mandla district which took possession of the road, and actually stopped the traffic. This was not the generally accepted specimen of a man-eater, old and mangy, but an exceedingly powerful beast of unexampled ferocity and audacity. It was a merciless highwayman, which infested a well-known portion of the road, and levied toll upon the drivers of the native carts, not by an attack upon their bullocks, but by seizing the driver himself, and carrying him off to be devoured in the neighbouring jungle. It had killed a number of people, and nothing would induce a native to venture upon that fatal road with a single cart; it had therefore become the custom to travel in company with several carts together, as numbers were supposed to afford additional security. This proved to be a vain expectation, as the tiger was in no way perplexed by the arrangement; it bounded from the jungle where it had lain in waiting, and having allowed the train of carts to pass in single file, it seized the driver of the hindmost, and as usual carried the man away, in spite of the cries of the affrighted companions.

Upon several occasions this terrible attack had been enacted, and the traffic was entirely stopped. A large reward was offered by the Government, but without effect; the man-eater never could be found by any of the shikaris.

At length the Superintendent of Police, Mr. Duff, who unfortunately had lost one arm by a gun accident, determined to make an effort at its destruction, and he adroitly arranged a plan that would be a fatal trap, and catch the tiger in its own snare. He obtained two covered carts, each drawn as usual by two bullocks. The leading cart was fitted in front and behind with strong bars of lashed bamboo, which formed an impervious cage; in this the driver was seated, while Mr. Duff himself sat with his face towards the rear, prepared to fire through the bars should the tiger, according to its custom, attack the driver of the rearmost cart. This would have been an exciting moment for the driver, but Mr. Duff had carefully prepared a dummy, dressed exactly to personate the usual native carter; the bullocks, being well trained, would follow closely in the rear of the leading cart, from which a splendid shot would be obtained should the tiger venture upon an attack.

All went well; the road was desolate, bordered by jungle upon one side, and wild grass-land upon the other. They had now reached the locality where the dreaded danger lay, and slowly the carts moved along the road in their usual apathetic manner. This must have been an exciting moment, and Mr. Duff was no doubt thoroughly on the lookout. Suddenly there was a roar; a large tiger bounded from the jungle, and with extraordinary quickness seized the dummy driver from his seat upon the rearmost cart, and dragged the unresisting victim towards the jungle!

Nothing could have been better planned, but one chance had been forgotten, which was necessary to success. No sooner had the tiger roared, and bounded upon the cart, than the affrighted bullocks, terrified by the dreadful sound, at once stampeded off the road, and went full gallop across country, followed by Mr. Duff's bullocks in the wildest panic. It was impossible to fire, and after a few seconds of desperate chariot race, both carts capsized among the numerous small nullahs of the broken ground, where bullocks and vehicles lay in superlative confusion; the victorious man-eater was left to enjoy rather a dry meal of a straw-stuffed carter, instead of a juicy native which he had expected.

This was a disappointment to all parties concerned, except the dummy driver, who was of course unmoved by the failure of the arrangement.

The story is thoroughly authenticated, and has been told to me by the Commissioner of the district exactly as I have described it. The tiger was subsequently killed by a native shikari, when watching from a tree over a tied buffalo.

Although the tiger as a "man-eater" is a terrible scourge, and frequently inflicts incredible loss upon the population of a district, there are tigers in existence which would never attack a human being, although they exist upon the cattle of the villages, and have every opportunity of seizing women and children in their immediate neighbourhood. About nine years ago there was a well-known animal of this character at a place called Bhundra in the Jubbulpur district, which was supposed to have killed upwards of 500 of the natives' cattle. This was a peculiarly large tiger, but so harmless to man that he was regarded merely in the light of a cattle-lifter, and neither woman nor child dreaded its appearance. The natives assured me that during fourteen years it had been the common object of pursuit, both by officers, civilians, and by their own shikaris, but as the tiger was possessed by the devil it was quite impossible to destroy it. This possession by an evil spirit is a common belief, and in this instance the people spoke of it as a matter of course that admitted of no argument; they assured me that the tiger was frequently met by the natives, and that it invariably passed them in a friendly manner without the slightest demonstration of hostility, but that it took away a cow or bullock in the most regular manner every fourth day. It varied its attentions, and having killed a few head of cattle belonging to one village, it would change the locality for a week or two, and, take toll from those within a radius of four or five miles, always returning to the same haunts, and occupying or laying up in the same jungle. The great peculiarity of this particular tiger consisted in the extreme contempt for fire-arms: it exposed itself almost without exception when driven by a line of beaters, and when shot at it simply escaped, only to reappear upon the following day. I was informed that everybody that had gone after it had obtained a shot, but bullets were of no use against a devil, therefore it was always missed.

I was 30 miles distant when I heard of this tiger, and I immediately directed our course towards Bhundra. It was a pretty and interesting place, where the presence of rich hematite iron ore has from time immemorial induced a settlement of smelters. There are jungle-covered low hills upon which large trees are growing, yet all such important mounds are composed of refuse from furnaces, which were worked some hundred years ago.

We arrived there early in May during the hottest season, and the clear stream below the village, rushing over a rocky bed, was a sufficient attraction to entice the animals from a great distance. This would account for the permanent residence of tigers.

The headman was a Thakur, a person of importance, and, as our camp had been sent forward on the previous day, we found everything in readiness upon our arrival; the Thakur and his people were in attendance.

After the usual salutations, I inquired concerning the celebrated tiger:
" How long was it since it had been heard of?"

The Thakur placidly inquired of our attendant, and I was informed that three days had elapsed since it killed the last cow; it would therefore in all probability kill another animal to-morrow. There was no excitement visible, but the natives spoke of the tiger as coolly and as unconcernedly as though it had been the postman.

My shikari was present, and I ordered him to tie up a good large buffalo, in prime condition, as the tiger was in the habit of selecting the best cattle for attack. After some delay, an excellent buffalo was brought for inspection, about sixteen months old, in fine condition, and there was little doubt that the tiger would attack, as the period had arrived when they might expect a kill.

The Thakur knew the exact position for the buffalo as bait, and he coolly assured me that the tiger would certainly kill, and that on the following day I should as certainly get a shot, but that the bullet would either fall from the hide, or in some way miss the object. He declared that upon several occasions he had himself obtained a shot, like everybody else, but it was useless, therefore he had long since ceased to take the trouble. This was rather interesting, and added to the excitement.

At daybreak on the following morning my eager shikari with several natives arrived, with news that the buffalo was killed and dragged into a dry bed of a rocky nullah within the jungle; and from the high bank they had seen the tiger devouring the hindquarters. This was satisfactory, although I was afraid that the tiger might have been disturbed by the inquisitiveness of the people; however, they laughed at the suggestion, and the beaters being ready, we sallied out to make a drive for a hopeless beast that was possessed by the devil.

The natives had been accustomed for so many years to act as beaters for this well-known animal that they had not the slightest nervousness; they knew the ground thoroughly, and the old mucharns, which had been vainly occupied so often, had simply been strengthened, but were ready in their original positions.

We had a large force of men, and several shikaris of long experience in the locality; it was accordingly a wise course to remain silent, as the people would have been confused by unnecessary orders.

Having left the line of men in position, we were taken about a mile in advance. I had given my shikari a double-barrelled gun, and I ordered him to take his stand as instructed by the natives; he accordingly disappeared, I knew not where. We entered the jungle, and presently descended the face of a small hill; then crossing a nullah, I was introduced to my mucharn; this was arranged upon a large tree which grew exactly upon the margin, and commanded not only the deep nullah beneath, but two other smaller nullahs which it met at right angles only a few paces distant. This looked well, as the tiger would probably slink along these secluded watercourses, in which case I should obtain a splendid shot. I climbed from the back of my steady elephant into the lofty perch; the people and animals left me to watch, squatted in a most uncomfortable position, as at that time I had not invented my charming turnstool.

At least an hour passed before I even heard the beaters. At length, amidst the cooing of countless doves, I detected the distant thud, thud of a tom-tom, and then the confused sound of many excited voices.

A few peacocks ran across the nullah; then a small jungle-sheep made the dead leaves rattle as it dashed wildly past; and almost immediately I heard a quick double shot about 200 yards upon my left.

I knew this must be my shikari, Sheik Jhan, and I felt sure that he had missed, as the two shots were in such rapid succession. If the first had struck the object, the second would not have been fired so quickly; if the first had missed, the exceeding quickness of the second shot would suggest confusion.

After waiting at least ten minutes without a sound of any animal, I whistled for the elephant, and descending from my post, I rode towards the position of Sheik Jhan.

A crowd of beaters were assembled, some of whom were engaged in searching for the bullets which he had fired, both of which had missed the tiger when within 12 yards' distance, although marching slowly over the sands and rocks in the bed of a large river; the natives were digging with pointed sticks into a grassy mound of sand.

Sheik Jhan described that an immense tiger had quietly passed close to him, but that no doubt it had a devil, as neither bullet had taken the least effect.

This was the customary termination; therefore no other course was left than to return to camp, the result having verified the prediction of the natives.

We now steered direct for the carcase of the buffalo, about 1 1/4 mile distant. Upon our arrival in the rocky bed of a dry river, where the smell of the tiger was extremely strong, we found the remains of the buffalo, a small portion of which had been eaten; I was assured by those who knew the habits of this tiger that it would return during the night, and that upon the following morning we should certainly obtain another shot.

I amused myself during the day by visiting the various smelting furnaces, all of which were upon a small scale, although numerous, and the method pursued was the same which I have found invariable among savage people. This consists in strong bellows worked by hand, the draught being sustained by continual relief of blowers, while the furnaces are constructed of clay, in the centre of which a small hole contains about a bushel of finely broken ore. Some powdered limestone was used as a flux, and the produce of a hard day's work, with five or six men employed, was about 15 lbs. of iron of the finest quality. This was never actually in a fluid molten state, but it was reduced when at white heat to a soft spongy mass resembling half-melted wax; it was then alternately hammered and again subjected to a white heat, until it arrived at the required degree of purity. The fuel was charcoal prepared from some special wood.

In the evening I pondered over the failure of Sheik Jhan, who declared that the tiger had taken him by surprise, as it had appeared while the beaters were so far distant that he could only just distinguish their voices. I came to the conclusion that this was the reason which explained the general escape of this wary animal, as it moved forward directly that the line of beaters entered the jungle, instead of advancing in the usual manner almost at the end of the beat. The sudden apparition of the tiger before it was expected would probably startle the gunner, who by firing in a hurry would in many instances entail a miss. Having well considered the matter, I determined to make myself more comfortable on the morrow, by padding the mucharn with the quilted pad of the riding elephant, and by sitting astride a tightly bound bundle of mats.

I would not allow any person to visit the carcase on the following morning, as I accepted the natives' assurance that the tiger would return to its kill; I gave orders that all beaters were to be in readiness, and we were to start together.

The morning arrived, and we started with a large force of nearly 200 men.

Upon approaching the spot where the carcase of the buffalo was left, I dismounted, and with only one man, I carefully inspected the position. The body had been dragged away. That was sufficient evidence, and I would not risk a disturbance of the jungle by advancing farther upon the tracks.

In order to maintain the most perfect silence, the beaters were kept at a considerable distance, and the line was to be formed only when a messenger should be sent back to say that the guns were already in position.

The native shikaris now assured me in the most positive manner that the tiger would certainly advance along the nullah, and would pass immediately beneath the tree upon which my mucharn of yesterday was placed.

Upon arrival at the tree I arranged the quilted pad and bundle of rugs in the mucharn, and having instructed my men to clear away a few overhanging creepers that in some places intercepted the line of sight along the nullah, I took my place, having carefully screened myself by intertwining a few green boughs to the height of 2 feet around my hiding-place.

I was comparatively in luxury upon the quilted mattress, and I waited with exemplary patience for the commencement of the beat in solitary quiet. A long time elapsed, as our messenger had to return about a mile before the line should receive orders to advance.

In the meanwhile I studied the ground minutely. I could see for 50 yards along the nullah, also there was a clear view where it joined the other approaches by which the tiger was expected. Exactly in front, on the other side the nullah beneath me, the jungle rose in a tolerably steep inclination upon a slope which continued for several hundred yards. If the tiger were to quit the nullah by which it would approach upon my left, it would probably cross over this hill to ensure a short cut, instead of continuing along the bottom of the nullah; this is frequently the habit of a tiger.

It was difficult to decide whether the beat had commenced, owing to the ceaseless cooing of the numerous doves, but presently a peacock flew into the tree upon my right, and almost immediately two peahens ran over the dead leaves, which made an exciting rustle in the quiet nullah. I felt sure that the beaters were advancing, as the peafowl were disturbed; I therefore kept in readiness, with rifle at full cock, as I felt sure that should the tiger exhibit himself, he would be far in advance of the approaching drive.

My ears were almost pricked with the strain of expectation, and I shortly heard the unmistakable beat of the native tom-tom.

Hardly had the sound impressed itself upon the ear, when a dull but heavy tread upon the brittle leaves which strewed the surface arrested my attention. This was repeated in so slow but regular a manner, that I felt sure it denoted the stealthy step of a tiger. I looked along the different nullahs, but could see nothing. The sound ceased for at least a minute, when once more the tread upon dead leaves decided me that the animal was somewhere not far distant. At this moment I raised my eyes from the nullahs in which he was expected, and I saw, through the intervening leafless mass of bushes upon the opposing slope, a dim outline of an enormous tiger, so indistinct that the figure resembled the fading appearance of a dissolving view. Slowly and stealthily the shadowy form advanced along the face of the slope, exactly crossing my line of sight. This was the "possessed of the devil" that had escaped during so many years, and I could not help thinking that, many persons would risk the shot in its present position, when the bullet must cut through a hundred twigs before it could reach the mark, and thus would probably be deflected. The tiger was now about 40 yards distant, and although the bushes were all leafless, there was one exception, which lay in the direct path the tiger was taking, a little upon my right; this was a very dense and large green bush called karoonda. Exactly to the right, upon the edge of this opaque screen, there was an open space about 9 or 10 feet wide, where a large rotten tree had been blown down; and should the tiger continue its present course it would pass the karoonda bush and cross over the clear opening. I resolved to wait; therefore, resting my left elbow upon my knee, I covered the shoulder of the unconscious tiger, and followed it with the .577 rifle carefully, resolved to exorcise the devil that had for so long protected it.

The shouts of the beaters were now heard distinctly, and the loud tom-tom sounded cheerfully as the line approached. Several times the tiger stopped, and turned its head to listen; then it disappeared from view behind the dense screen of the karoonda bush.

I lowered the rifle, to rest my arm for a moment. So long a time elapsed, that I was afraid the tiger had turned straight up the hill in a direct line with the bush, and thus lost to sight; I had almost come to this sad conclusion, when a magnificent head projected from the dark green bush into the bright light of the open space. For quite 15 seconds the animal thus stood with only the head exposed to view, turned half-way round to listen. I felt quite sure that I could have put a bullet through its brain; but I waited. Presently it emerged, a splendid form, and walked slowly across the open space. At the same moment as I touched the trigger, the tiger reared to its full height upon its hind legs, and with a roar that could have been heard at a couple of miles' distance it seized a small tree within its jaws, and then fell backwards; it gave one roll down the slope, and lay motionless. The devil was cast out.

I never saw such enthusiastic rejoicing as was occasioned by the death of this notorious tiger. The news ran like fire through the neighbouring villages before we had completed the packing of the animal upon Demoiselle. I had no means of weighing this tiger, but it was the heaviest I have ever seen, and although we had four poles beneath its body and a great number of willing men at the extremities, we had great difficulty in loading Demoiselle. By the time we had completed the operation we had a large crowd in attendance, all of whom followed the elephant upon the march towards our camp bearing the body of the tiger, which had been the scourge of their herds during so many years.

At least 300 women and children assembled to satisfy themselves that their enemy was really dead. The women kissed his feet and wiped their eyes with the tip of his tail; for what purpose could not be explained.

As this animal had lived in luxury, it was immensely fat, and we filled numerous chatties with this much-loved grease, to be used as ointment for rheumatic complaints. Unfortunately at that time I had no weighing machine, therefore it was impossible to judge the weight with accuracy, but we computed that the fat alone amounted to 70 lbs. avoirdupois. The tiger was certainly upwards of 500 lbs.

I found the '577 bullet of pure lead had entered exactly at the shoulder joint, which it had smashed to atoms, carrying splinters of bone through the lungs; passing through the ribs upon the opposite side, it had smashed the left shoulder, and was fixed beneath the skin, expanded like a mushroom.

There was no danger to any person employed in this hunt, but I have described it as an apt example of a cunning tiger, which escaped so many attempts upon its life that it was regarded as "uncanny."

My servant Thomas was quite delighted, as he had offered to bet that, "devil or no devil, his master's rifle would kill him, if he got a shot."

It has been generally admitted that the great variety of this species renders a classification almost impossible. Different countries adopt special names for the varieties which inhabit the localities; the leopard may be termed a panther, or cheetah, or wild cat, or even a jaguar, but it remains a leopard, differing in size, colour, and form of spots, but nevertheless a leopard. I shall therefore accept that name as including every variety. Although the genus Felis embraces in its nomenclature all the various representatives, from the lion (*Felis Leo*) to the ordinary domestic cat, the two principal examples of the race, the lion and tiger, are totally distinct from all others in their natural characters. The leopard is far more daring; at the same time it is infinitely more cautious, and difficult to discover.

No lion or tiger can ascend a tree unless the branches spring from within 4 or 5 feet of the ground; even then it would be contrary to the habits of the animal to attempt an ascent, although it might be possible under such favourable circumstances. A leopard will spring up a smooth-barked tree with the agility of a monkey; and there is a small species which almost lives among the branches (F. Macroscelis), from which it leaps upon its prey when passing unconsciously beneath.

An examination of the skins of leopards from various portions of the globe exhibits a striking difference in colouring and quality of fur. We find the snow leopard, which inhabits the Himalayahs and other lofty mountain ranges, with a fur of great value, deep and exceedingly close, while the spots are not determined as distinct black, but are shaded off by gray. This species is generally found at altitudes of from 8000 to 10,000 feet, or even higher. In Manchuria and the Corea there is a species which is unknown in India; this is a large animal, with a peculiarly rich and deep fur when killed during winter; the black spots are exceedingly large, and are formed in rings. A skin in my possession measures 7 ft. 9 in. in length; the tail is full, and the fur long; this is unusually beautiful, and it must have inhabited some lofty altitude where the temperature was generally moderate.

In Africa the leopards have almost invariably solid black spots, very close together upon the back, and becoming less crowded towards the belly and flanks. In Ceylon there are two distinct varieties-the large panther, generally about 7 ft. 6 in. in length, and a smaller leopard, which inhabits the mountains; in that island of misnomers they are both included in the name cheetah.

In India there are several varieties, and the largest is generally distinguished as a panther. There is no animal more commonly distributed in the world than the leopard, and no tropical country is free from this universal pest, unless an island formation has excluded its unwelcome presence.

It is difficult to determine the limit in the gradation of size at which this animal merges from the leopard into the wild cat. The varieties of cats are so numerous that I do not pretend to describe them; some are of sufficient importance to be classed among the smaller leopards, while others are no larger than the ordinary domestic cat. These vary through every shade of feline colouring, from spots to stripes, or to a fulvous brown similar to the tawny coat of a lioness; but, notwithstanding the difference in shades and spots, in cats and in the true leopard or panther the character is the same. They are all cunning, ferocious, and destructive, and I believe that far more cattle and goats are killed by leopards throughout the Indian Empire than by the usually accredited malefactor, the tiger.

The largest and most beautifully marked of the leopards is the jaguar of South America. This is the size of a small tigress, and is more heavily framed than any of the leopards; the head is especially large, and the animal might almost be termed a spotted tiger. The rings are peculiarly marked, and waved instead of being circular.

The cheetah or hunting leopard is a distinct species, and although classed among the leopards, it is altogether different, both in habits and appearance; the claws, although rather long, are not retractile, neither are they curved to the same extent as all others of the genus Felis, but they resemble somewhat the toe-nails of the dog. I shall accordingly separate this animal from the ordinary class of leopards, and give it a separate existence as an object of natural history.

The panther or larger variety of leopard is about 7 ft. 6 in. in length, and has been known to approach closely upon 8 feet, but this would be an unusual size. This animal is exceedingly powerful, with massive neck and strongly developed legs. The weight of a fine specimen would be from about 160 lbs. to 170 lbs. Although heavy, there is no animal more active, except the monkey, and even those wide-awake creatures are sometimes caught, by the ever-watchful panther. Stories are told of accidents that have occurred when the hunter has been pulled out of his tree, from which imaginary security he was watching for his expected game. It is impossible to deny such facts, although they are fortunately rare exceptions to the general rule; but there can be no doubt that a panther or leopard would attack upon many occasions when a tiger would prefer to slink away.

The habits of the leopard are invariably the same, it prowls stealthily about sunset and throughout the night in search of prey. It seizes by the throat and clings with tenacious claws to the animal's neck, until it succeeds either in breaking the spine, or in strangling its victim, should the bone resist its strength. When the animal is dead, the leopard never attacks the hind-quarters first, according to the custom of the tiger, but it tears the belly open, and drags out all the viscera, making its first meal upon the heart, lungs, liver, and the inside generally. It then retreats to some neighbouring hiding-place, and, if undisturbed, it will return to its prey a little after sundown on the following day.

It is far more difficult to circumvent a leopard than a tiger; the latter seldom or never looks upwards to the trees, therefore it does not perceive the hidden danger when the hunter is watching from his elevated post; but the leopard approaches its kill in the most wary and cautious manner, crouching occasionally, and examining every yard of the ground before it, at the same time scanning the overhanging boughs, which it so frequently seeks as a place of refuge.

Upon many occasions, when the disappointed watcher imagines that the leopard has forsaken its kill, and that his patience will be unrewarded, the animal may be closely scanning him from the dense bush, under cover of which it was noiselessly approaching. In such a case the leopard would retreat as silently as it had advanced, and the watcher would return home from a fruitless vigil, under the impression that the leopard had never been within a mile of his position. One of the cleverest birds in creation is the ordinary crow of all tropical countries, which lives well by the exercise of its wits; nothing escapes the observation of this bird, and it is the first to discover the body of any animal that may have been killed. Should one or more of these birds be perched in the trees after sunset, near the carcase of an animal, and should it utter a "caw," when at that late hour it should have gone to roost, you may be assured that it has espied an approaching leopard, although it may be invisible to your own sight. The watcher should be careful not to move, but to redouble his vigilance in keeping a bright look-out, as the leopard will be equally upon its guard should it hear the cry of the warning crow.

There is very little sport afforded by this stealthy animal, and it is almost useless to organize a special hunt, as it is impossible to form any correct opinion respecting its locality after it has killed an animal. It may either be asleep in some distant ravine, or among the giant branches of some old tree, or beneath the rocks in some adjacent hill, or retired within a cave, but it has no special character or custom that would guide the hunter in arranging a beat according to the usual rules in the case of tigers. The leopard is merely a nuisance, and as such it should be treated as vermin, and exterminated if possible.

There are various forms of traps adopted by the natives in different countries; the most certain is the old-fashioned fall, similar upon a large scale to the common fall mouse-traps. These should be permanent fixtures in various portions of the jungles, and they should be baited whenever the tracks of a leopard may be discovered in the neighbourhood. The trap is formed by an oblong 10 feet by 3 of very strong and straight palisades, sunk 2 feet deep in the ground, and well pounded in with stones. These should be 5 feet high, with a fall door at one end. The top should be closely secured with heavy cross-pieces of parallel logs, well weighted with big stones.

The rear of this trap should be partitioned with bamboo cross-bars to form a cage, in which either a goat or a village dog should be tied as a living bait. Leopards are particularly fond of dogs, and the advantage of such a bait during the night consists in the certainty that the dog, finding itself alone in a strange place, will howl or bark, and thereby attract the leopard. The partition must be made of sufficient strength to protect the animal from attack. In Africa the natives form a trap by supporting the fallen trunk of a large tree in such a manner that it falls upon the leopard as it passes beneath to reach the bait. This is very effective in crushing the animal, but it is exceedingly dangerous, like all other African traps, as it would kill any person or other creature that should attempt to pass. Newera Ellia, the mountain sanatorium of Ceylon, was always well furnished with leopard-traps upon the permanent system, and the leopards, which were at one time a scourge of the neighbourhood, were considerably reduced. In 1846 I introduced English breeds of cattle and sheep, and started an agricultural settlement at that delightful mountain refuge from tropical heat; but the leopard became our greatest enemy, and although the cattle were well housed at night, and carefully watched when at pasture during the day, our losses were severe. I observed a peculiarity in the attacks by leopards; they seldom appeared upon a bright summer day, but during the rainy season, when the wind was howling across the plain, and driving the cold mist and rain, the cattle were off their guard, and generally turned their tails to the chilly blast. It was invariably during such weather that the leopards attacked. The watchman was probably wrapped in his blanket, wet, and shivering beneath a tree, instead of remaining on the alert, and this auspicious moment was selected by the leopard for a successful stalk upon the unsuspecting herd. I have frequently lost both cows and sheep, that were attacked and killed in broad daylight, and the leopards were generally of sufficient strength to break the neck of a full-grown beast. It should be remembered that the native cattle are much smaller than those of Europe, and I do not think it would be possible for a leopard to dislocate the neck of any English cow. An example occurred when unfortunately a valuable Ayrshire cow was attacked, and the leopard completely failed in the usual dexterous wrench, but the throat was so mangled that the cow died within a few days, although the leopard was driven away by the watchman almost immediately upon its onset.

The wounds from the claws of a leopard are exceedingly dangerous, as the animal is in the habit of feeding upon carcases some days after they have been killed; the flesh is at that time in an incipient stage of decomposition, and the claws, which are used to hold the flesh while it is torn by the teeth and jaws, become tainted and poisoned sufficiently to ensure gangrene by inoculation. The claws of all carnivora are five upon each of the fore feet, including the useful dew-claw, which is used as a thumb, and thoroughly secures the morsel while the animal is pulling and tearing away the muscles from the bones.

A wound from either a tiger or a leopard should be thoroughly syringed with cold water mixed with 1/35th part of carbolic acid, and this syringing process should be continued three times a day whenever the wound is dressed. Nothing should be done but to wrap the wound with linen rag soaked in the same solution, and keep it continually wetted.

The daring of a leopard during night is extraordinary. I have frequently during wet weather discovered in the early morning a regular beaten track in the soft earth, where a leopard has been prowling round and round a cattle-shed containing a herd of animals, vainly seeking for an entrance.

At one time my own blacksmith had a nocturnal adventure with a leopard which afforded a striking example of audacity. A native cow had a calf; this being her first-born, the mother was exceedingly vicious, and it was unsafe for a stranger to approach her, especially as her horns were unusually long, and pointed. The cattle-shed was scarped out of the hillside, and was within a few feet of the blacksmith's house. The roof was thatched. During the night, a leopard, which smelt the presence of the cow and calf, mounted the roof of the shed and proceeded to force an entrance by scratching through the thatch. The cow at the same time had detected the presence of the leopard, and, ever mindful of her calf, she stood ready to receive the intruder, with her sharp horns prepared for its appearance. It is supposed that upon the leopard's descent it was at once pinned to the ground, before it had time to make its spring.

The noise of a tremendous struggle aroused the blacksmith, who, with a lantern in his hand, opened the cattle-shed door and discovered the cow in a frantic state of rage, butting and tossing some large object to and fro, which evidently had lost all power of resistance. This was the leopard in the last gasp, having been run through the body by the ready horns of the courageous mother, whose little calf was nestled in a corner, unmindful of the maternal struggle.

No sooner had the blacksmith appeared upon the scene, than the character of the conflict changed, and the cow, regarding him in the light of a fresh enemy, left the crumpled body of her antagonist and charged straight at her proprietor, who dropped his lantern and flew to the arms of his wife, whom he had left in bed. After some delay, during which the courage of all parties was restored, excepting that of the crippled leopard, the cow was appeased, and a shot from a pistol through the head of the enemy closed the episode.

Every resident in India is aware of the depredations committed by this pestilent class of the carnivora. Lions and tigers may be dangerous in the jungles in every country which they inhabit, but they never invade the actual premises; it is exactly there where the leopard is to be feared. Nothing is too small or too large for its attack; from a fowl upon the roost to a cow in the pasturage, all that belongs to the domestic stock is fair game for the wily leopard.

The cautious approach of this animal is so wary that a dog is pinned by the neck and carried off before it is aware of the presence of its enemy. Upon one occasion in Africa we were bivouacked for the night on the banks of the Settite river, and no sound disturbed the repose of the camp. Suddenly a leopard bounded into the centre, where the Arabs were sleeping around the embers of a splendid fire, and seizing one of the dogs, it sprang into the darkness, carrying its captive with it. The remaining dogs rushed off in pursuit, together with all the Arabs with swords and shields, and the leopard dropped its prize about 150 yards from our enclosure. The unfortunate dog had been surprised in its sleep, and it died in a few hours from the injuries sustained, the neck and throat being terribly lacerated. It would have been natural to suppose that the dogs would have given an alarm on the approach of the wild animal, but the noiseless tread of the leopard, as usual, was unheard, even in the extreme stillness of a calm night. The sudden attack of a leopard is generally so unexpected that a dog has no time for self-defence, and being invariably seized by the neck, it is at once rendered helpless, and cannot utter a warning shriek before it is carried off. I was walking with a very powerful bull terrier at Newera Ellia in Ceylon, when the dog, who was running through the jungle within a few yards of me, suddenly disappeared without a cry, and was never heard of again; this same dog would have made a good defence had it confronted the leopard face to face.

On another occasion a dog named Matchless, a cross between foxhound and pointer, was seized by a leopard in open day when, together with a pack of hounds, walking through a jungle-path at Dimbola, not far from Newera Ellia. The leopard sprang suddenly from a tree, and, seizing the dog, immediately ascended, and took refuge among the boughs with the hound suspended in its mouth. The entire pack bayed the audacious enemy; it then dropped the dog and jumped from tree to tree, followed beneath by the excited hounds. At length the leopard reached a large tree, which was sufficiently isolated to prevent it from springing to any adjoining branches. In this position it was surrounded, and became the central object, where it remained snarling at the infuriated pack. The party of hunters now commenced a bombardment with stones, and a lucky hit induced the leopard to either jump or fall into the middle of the hounds. There was an exceedingly large dog named Pirate, a cross between mastiff and bloodhound; he immediately seized the leopard, and a general fight ensued, the whole pack supporting Pirate in his attack. Captain E. Palliser, late 7th Hussars, quickly thrust his hunting-knife under the shoulder, and in a few minutes the hounds were worrying a dead leopard.

Some few years ago the hounds belonging to the late Mr. Downall hunted a leopard at Newera Ellia, and a tremendous struggle ensued. There were several very powerful and large seizers among the pack, and the enemy was overmatched, but although the big dogs had the mastery of the animal, they could not actually kill it outright. General J. Wilkinson was on the spot, and he thrust his hunting-knife into the fatal spot; but he was a little to slow in withdrawing the blade; the dying leopard made a quick blow with its fore paw, and inflicted a serious wound upon his hand, lacerating the muscles of the thumb to a degree that rendered surgical treatment necessary for several weeks. When using the hunting-knife, extreme dexterity is to be observed in delivering the stab, and instantaneously recovering the weapon. There is no object to be gained by keeping the knife within the wound, and there is considerable danger of injury to the hand. If the knife is used by an expert it will never be held with the point downwards like a dagger, but the handle will be grasped for a direct thrust, as though the weapon were a sword. In this position the knife is always well under command, and it can be instantly withdrawn and the thrust repeated upon a favourable opportunity.

I had a very savage and powerful dog many years ago which was a cross of Manilla bloodhound with some big bitch at the Cape of Good Hope. This animal weighed upwards of 130 lbs., and became a well-known character in the pack, which I kept for seven years in Ceylon. Although I never actually witnessed a duel between this dog and a leopard, such an event frequently took place. It was the custom of Smut to decline all control, and when the hounds were secured in couples to prevent them from following the scent of a leopard, should recent tracks be visible in the jungle, this determined dog would erect the bristles on his back, emit low growls when summoned back, and would disappear to hunt up, single-handed, the scent of the dreaded enemy. Upon these occasions Smut would be unheard of during the remainder of the day, and he would return to kennel in the evening, proudly trotting along, covered with blood and wounds, but always so fierce that he refused all aid and medical attendance; he was merely ready for his dinner. He had of course tackled his adversary, and indulged his propensity for a stand-up fight, with results which we never could discover; probably the leopard had been glad to retire honourably from the uncertain conflict. This grand dog was ultimately killed in a fight with an immense boar, and his name will reappear in connection with the sambur deer, misnamed the "elk," throughout Ceylon.

It is most discouraging to lose good dogs through the stealthy attacks of leopards, and in looking back to the list of casualties among
the pack when I kept hounds in Ceylon it is distressing to see the
number which were taken by these unsparing animals. If a hound is
lost in the jungle, it will certainly sit down and howl, thereby exhibiting considerable intelligence, as it is, in fact, crying for assistance;
but such a cry will attract the ever-wary leopard, who will probably
approach by leaping from tree to tree, and pounce upon the unfortunate dog before it is aware of the impending danger. The hound
that would have offered a stout resistance if boldly attacked face to
face, has no more chance than an Irish landlord when shot at by an
assassin secreted behind a wall by the roadside.

This noiseless approach may be imagined from an incident which
occurred to me in Abyssinia, when watching a pool by moonlight,
in a deep bend of the river Royan during the dry season; all streams
had evaporated, excepting an occasional deep hole in a sudden
curve of the exhausted bed. Hours had been passed, but nothing
larger than antelopes had appeared. We were sitting beneath a very
large tree completely denuded of leaves, and the moon was shining
brightly, producing a sharp outline of every bough. Suddenly my
wife pulled my sleeve and directed my attention to a large animal
crouched upon the branches exactly above us. I might have taken a
splendid shot, but I at first imagined it to be a dog-faced baboon
(Cynocephalus) that had been asleep upon the tree. I stood erect to
obtain a clearer view, and at once the object sprang to the ground
within a few feet of us and bounded into the jungle. This was a
leopard, which had probably reached the tree by means of some
neighbouring branch, and so noiselessly that we had not discovered
its presence. The animal had evidently winded us, and determined
to reconnoitre our position.

In every country the natives are unanimous in declaring that the
leopard is more dangerous than the lion or tiger, and I quite agree
in their theory that when any dangerous animal is met with, the
traveller should endeavour to avoid its direct gaze. It is an error to
suppose that the steady look from the human eye will affect an
animal by a superior power, and thereby exert a subduing influence; on the contrary, I believe that the mere fact of this concent-
ration of a fixed stare upon the responding eyes of a savage animal

will increase its rage and incite attack. If an animal sees you, and it imagines that it is itself unobserved, it will frequently pass by, or otherwise retreat, as it believes that it is unseen, and therefore it has no immediate dread; but if it is convinced that you mean mischief, by staring it out of countenance, it will in all probability take the initiative and forestall the anticipated attack.

A leopard will frequently attack if it is certain that your eyes have met, and it is always advisable, if you are unarmed, to pretend to disregard it, at the same time that you keep an acute look-out lest it should approach you from behind. Wherever I have been in Africa, the natives have declared that they had no fear of a lion, provided that they were not hunting, as it would certainly not attack them unprovoked; but that a leopard was never to be trusted, especially should it feel that it was discovered. I remember an occasion when the dry grass had been fired, and a native boy, accompanied by his grown-up brother, was busily employed with others in igniting the yellow reeds on the opposite bank of a small stream, which had checked the advance of the approaching flames. Being thirsty and hot, the boy stooped down to drink, and he was immediately seized by a leopard, which sprang from the high grass. His brother, with admirable aim, hurled his spear at the leopard while the boy was in its jaws; the point separated the vertebrae of the neck, and the fierce brute fell stone dead. The boy was carried to my hut, but there was no chance of recovery, as the fangs had torn open his chest and injured the lungs; these were exposed to view through the cavity between his ribs. He died during the night. The muscular strength of the jaws and neck is very marked in all the carnivora, and the skull when cleaned is most disappointing, and insignificant if compared with the size of a living head. This is especially the case with leopards, and it is difficult to believe that so small a pair of jaws can inflict a deadly wound almost immediately.

I have already remarked upon the wide difference in the size of leopards, showing that the largest, which are sometimes known as panthers, are almost equal to a small tigress. Some of this class possess extraordinary power, in carrying a heavy weight within their jaws. At a place called Soonbarro, in the Jubbulpur district, we were camped upon a large open space entirely devoid of bush. The ground was free from grass, and dusty, therefore the surface would expose every track. Three full-grown sheep were tied to the cook's tent, well secured to a strong peg. In the morning only two remained, but the large tracks of a leopard or panther were deeply printed in the dust, and the sheep had been carried off bodily, as a big dog would carry a hare. The jungle at the base of a range of hills, almost perpendicular and full of caves, was the great resort of leopards, bears, and jackals; the sheep had been actually carried quite half a mile without leaving a trace upon the ground to show that it had been partially dragged, or that the leopard had stopped to rest. This was an admirable proof of a great carrying power, as nothing could have moved upon that dusty surface without leaving a well-printed trace.

Although the cubs of leopards are charming playthings, and exhibit much intelligence and apparent affection, it is a great mistake to adopt such companions, whose hereditary instincts are certain to become developed in full-grown life and lead to grave disaster. The common domestic cat is somewhat uncertain with her claws, and most people must have observed that should they be themselves spared the infliction of a feline scratch, the seats and backs of morocco chairs are well marked by the sharp talons, which cannot refrain from exercising their power upon any substance that tempts the operation. I remember a leopard in Khartoum that was considered tame; this beast broke its chain, and instead of enjoying its liberty in a peaceful manner, it at once fastened upon the throat of a much-prized cow, and would have killed the animal had it not been itself beaten to death with clubs by a number of stout slaves of the establishment. All such creatures are untrustworthy, and they should be avoided as domestic pets. The only class of leopard that should become the companion of man is the most interesting of the species: this is the hunting leopard (Felis jubata). I have never met a person who has shot one of this species in a wild state, and such an animal is rarely met with in the jungle. Most people are under the impression that the hunting leopard with non-retractile claws is incapable of climbing a tree; I was myself of this opinion until I actually witnessed the act, and the animal ran up a tree with apparent ease, ascending to the top.

The Felis jubata is totally different in shape from all other leopards. Instead of being low and long, with short but massive legs, it stands extremely high; the neck is long, the head small, the eyes large and piercing; the legs are long, and the body light. The tail is extremely long, and thick; this appears to assist it when turning sharply at full speed. The black spots upon the skin are very numerous, and are simply small dots of extreme black, without a resemblance of rings. It is generally admitted that the hunting leopard is the fastest animal in the world, as it can overtake upon open ground the well-known black-buck, which surpasses in speed the highest bred English greyhound. I have never had experience of this animal in a wild state; those I have known were as gentle as dogs. It is a common mistake to suppose that they invariably approach their game by a stealthy stalk, followed by a few tremendous bounds, only to slink back if disgraced by defeat. I have seen them run a long course in the open, exactly like a greyhound, although the pace and action have resembled the long swinging gallop of a monkey. The nature of this beautiful creature is entirely opposed to the cat-like crouching tactics of the ordinary leopard: its large and prominent eyes embrace a wide field of view; the length of neck and legs, combined with the erect attitude of the head, denotes the character of the animal, as it includes a vast distance in its gaze, showing that it seeks its game upon a wide expanse of plain, instead of surprising the prey by an unexpected and treacherous attack. This is the only species that is a useful companion to man when engaged in field sports; and the native princes of India have from time immemorial been accustomed to train the Felis jubata for hunting deer and antelopes, precisely as European nations have adopted the greyhound for the coursing of hares.

The Guikwar of Baroda possesses first-class hunting leopards, and I had an opportunity of witnessing many good hunts when enjoying his hospitality at Dubka in 1880. The whole of that country is rich alluvial soil, which produces vast agricultural wealth. The fields are divided by exceedingly thin live fences formed by a species of Euphorbia; the country being flat, it affords the perfection of ground for riding, therefore such sport as pig-sticking or coursing may be enjoyed to the fullest extent. During our visit the Guikwar had most kindly arranged every kind and style of sport, including a pack of hounds, half a dozen well-trained cheetahs (hunting leopards), and a posse of hawks and falcons with their numerous attendants. The position of Dubka was supposed to be most favourable for a hunting centre, about 18 miles from the capital Baroda. There was a large palace for the Guikwar, and a convenient bungalow for his friends, situated about 30 yards from the cliff, which, 100 feet above the stream, commanded an imposing view of the river; this flowed beneath, about 3/4 mile in width during floodtime, but was now reduced to 300 or 400 yards in the dry season. A few miles from the bungalow there was a magnificent country for the cheetahs, as the ground, having been subject to inundations, was now perfectly dry, and exposed a large plain, like an open race-course, upon which the young grass was about 2 inches high. In the neighbourhood of this plain there were a few low hills covered with sparse jungle, and for several miles around, the flat surface was more or less overgrown with bush, interspersed with patches of cultivation.

On the first day's journey we travelled along a dusty road, which had never been metalled, for the reason that no stone existed in the neighbourhood; the wheels of the carriages sank deeply in the sandy loam, and the saddle was a far more enjoyable seat than a struggling wheeled conveyance. The falconers enlivened the journey by several flights at herons and cranes, which were very numerous in the marshes that bordered occasional lakes or jheels. We had the opportunity of observing the sagacity of a peregrine falcon, which, immediately upon being unmasked, rose straight in the air, instead of following the heron on its direct course. At first I imagined that it did not see the bird, which flew very high, and kept above the lake. Presently the falcon took a totally opposite direction, soaring to an altitude that reduced it to a mere speck. By this time the heron had cleared the large expanse of water, and was at a great height, perpendicular with the dry land beneath. The falcon made a sudden swoop, and with the velocity of a meteor it shot downwards upon an oblique course towards the unlucky heron. This bird had evidently been watching the impending danger, and it attempted to evade the attack by rising rapidly in the air, in order to destroy the advantage which a higher altitude had conferred upon the enemy. It was too slow: the falcon shot like an arrow to the mark, and struck the heron with such force that for the moment both birds, hanging together, fell for about 100 feet, as though hit by a rifle bullet. After the first blow, the large wings of the heron expanded, and checked the rapid fall; the falcon was fixed upon its back, holding the neck in its sharp beak, while it clung to the body with its claws. In this position the two birds slowly descended towards the ground, twirling round and round in their descent from a height of about 1000 feet.

In the meantime the falconers had been galloping at full speed around the lake, towards the spot upon which they had expected the birds to fall. The falcon was very savage, and it continued to tear the neck of the heron even when captured by the men. This was a cruel exhibition, as the head falconer, having taken possession of the birds, brought them to be admired, the heron being still alive, while the peregrine was tearing at its bleeding neck. He appeared surprised that I insisted upon its being killed, and he at once replaced the hood upon the falcon and prepared for another flight. He explained the reason for the peculiar behaviour of the falcon in taking a different direction from its game; it was afraid of the water beneath, into which both birds must have fallen had the heron been struck before it had cleared the surface; it had therefore attained a high altitude in a different direction, from which it could swoop obliquely when the lake no longer lay beneath them. This man was a high authority, and he assured me that many well-trained falcons would decline to strike a bird when flying across water, as they thoroughly understood the danger.

We had several good flights, in one of which a large crane succumbed after a very severe struggle, which seemed to test the utmost strength of the peregrine, but in every case the attack was delivered from a superior altitude, which left no chance of escape to the bird beneath; the result depended upon the power of the falcon to continue its hold during the struggles of the heavier and more powerful bird.

On the day following our arrival at Dubka, we devoted ourselves to hunting the black-buck with cheetah. In this sport, all persons, excepting the keepers of the animals, are simply spectators, and no interference is permitted. Each cheetah occupies a peculiar cage, which forms the body of a cart, drawn by two bullocks. When game is expected, the cheetah is taken from the cage, and occupies the outside seat upon the top, together with the keeper. The animal is blinded by a hood, similar to that worn by the falcon, and it sits upright like a dog, with the master's arm around it, waiting to be released from the hood, which it fully understands is the signal that game is sighted.

There were plenty of black-buck, and we were not long in finding a herd, in which were several good old buck, as black as night. Nothing could be more favourable than the character of the ground, for the natural habits of the cheetah. The surface was quite flat and firm, being a succession of glades more or less open, surrounded by scattered bush. A cheetah was now taken from its cage, and it at once leapt to the top, and sat with its master, who had released it from the hood. After an advance of about 200 yards, the wheels making no noise upon the level surface, we espied the herd of about twenty antelopes, and the cart at once halted until they had slowly moved from view. Again the cart moved forward for 70 or 80 paces, and two bucks were seen trotting away to the left, as they had caught a glimpse of the approaching cart. In an instant the cheetah was loosed; for a moment it hesitated, and then bounded forward, although the two bucks had disappeared. We now observed that the cheetah not only slackened its pace, but it crept cautiously forward, as though looking for the lost game.

We followed quietly upon horseback, and in a few seconds we saw the two bucks about 120 yards distant, standing with their attention fixed upon us. At the same instant the cheetah dashed forward with an extraordinary rush; the two bucks, at the sight of their dreaded enemy, bounded away at their usual speed, with the cheetah following, until all animals were lost to view among the scattered bushes.

We galloped forward in the direction they had taken, and in less than 300 yards we arrived at the spot where the cheetah had pinned the buck; this was lying upon its back without a struggle, while the firm jaws of the pursuer gripped its throat.

The cheetah did not attempt to shake or tear the prey, but simply retained its hold, thus strangling the victim, which had ceased all resistance.

The keeper now arranged the hood upon the cheetah's head, thus masking the eyes, which were gleaming with wild excitement, but it in no way relaxed its grip. Taking a strong cord, the keeper now passed it several times around the neck of the buck, while it was still held in the jaws of the cheetah, and drawing the cord tight, he carefully cut the throat close to the teeth of the tenacious animal. As the blood spurted from the wound, it was caught in a large but shallow wooden bowl or ladle, furnished with a handle. When this was nearly full, the mask was taken off the cheetah, and upon seeing the spoon full of blood it relaxed its grasp and immediately began to lap the blood from its well-known ladle. When the meal was finished, the mask or hood was replaced, and the cheetah was once more confined within its cage, as it would not run again during that day.

The wooden ladle is, to the cheetah, an attraction corresponding to the "lure" of a falcon; the latter is an arrangement of feathers to imitate a bird. The ladle is known by the cheetah to be always connected with blood, which it receives as a reward after a successful hunt; therefore, when loose, and perhaps disobedient to a call, it will generally be recovered by exhibiting the much-loved spoon, to which it returns, like a horse to a sieve of oats.

We now uncarted a fresh cheetah, and were not kept long waiting before we came upon a lot of antelopes, most of which were females and young bucks. At length, after careful stalking by driving the bullock-cart in an opposite direction to the herd, and then slightly turning to the left, in the endeavour to decrease our distance, we saw a fine buck standing alone within 100 yards, as we had not been observed while advancing through the scattered bush.

The cheetah lost not a moment, but springing lightly to the ground, it was at full speed, and within 50 yards before the unwary buck perceived it. Taken by surprise, instead of bounding off in mad retreat, this gallant little buck lowered its sharp-pointed horns and stood on the defence against the onset of its fierce antagonist. This was a pretty but a pitiable sight, as I knew that the odds were terribly against the buck; but in another instant the actual encounter took place, and I was surprised to see how well the plucky buck conducted the defence. It actually charged the advancing cheetah, and stopped its rush. The cheetah held back, and again the buck rushed in; but as we advanced, the poor little beast was evidently frightened at the people, and it turned to run. The moment that the cheetah saw its opportunity, it sprang forward; we saw the blow of the paw, delivered as quick as lightning upon the right haunch, and the gallant little buck was on its back, with its throat hopelessly throttled in the cheetah's jaws.

We were sorry for this termination, as I should like to have witnessed the result, had we not disturbed the fight by our presence. The keepers did not regard the affair in the same light, as they declared the cheetah might have been injured severely by the horns, but that eventually it would have killed the black-buck.

In a couple of days we had killed a number of these beautiful animals, but I became tired of the sport, as the affair was invariably over in a couple of minutes. One thing was certain, the cheetahs were first-rate, and there was none of the skulking and slinking back, which I had read of as characteristic of the hunting leopard.

This style of hunting must naturally depend upon the condition of the ground. We had hunted the localities that were in favour of the cheetah, when scattered bush admitted of a tolerably close approach; but after a couple of days we had scared the black-buck to such a degree that they entirely forsook the sparse covert, and took to the bare open plain, where it was simply impossible to approach them unobserved. This intensified the pleasure, as hitherto the cheetahs had triumphed in almost every hunt.

I accordingly suggested that we should confine our party to three mounted persons and three carts, with of course the same number of cheetahs, and endeavour to obtain some real coursing upon the open plain.

We started. There was hardly a bush upon the wide expanse of level ground, as smooth as a billiard table; only two or three trees occupied this large area, and they were unhealthy specimens, which looked as though periodical inundations had disagreed with them. We arrived upon this great natural race-course, and the binoculars were at once in request to scan the distant surface in search of the desired game. In a short time, as we advanced leisurely, constantly halting to take an observation, we discovered a considerable herd of about thirty or forty antelopes, among which there were two bucks perfectly black; these were feeding upon the short young grass in the very centre of the open ground. The question arose, "How in the world shall we get near them?" It was determined that our three horses should as much as possible conceal themselves on the right side of the three carts, and that they should attempt the approach by moving in a circle, getting nearer and nearer to the herd, as the black-buck family might become less shy, and more accustomed to the appearance of the carts. This plan was cleverly carried out by the drivers, and in about twenty minutes we had, by circling and alternately advancing direct, got to within 300 yards' distance. The herd was all together, as several times they had stopped feeding to gaze at our party, after which they had trotted off a little distance, and then closed up, as though for mutual protection, which gave confidence. We again halted, to try the effect upon the herd. They merely looked up, and for the moment ceased feeding, but almost immediately one of the bucks made an unprovoked attack upon the other, apparently with the intention of driving it away from the females. Instead of retreating from the insult, the affronted buck at once returned to the encounter, and a tremendous fight was the immediate result, the two combatants charging each other like rams, and boring, first one, then the other backward, with the greatest fury. During this duel the herd of females stood entranced, as admiring spectators of the struggle.

Not so our drivers, who, instead of their hitherto wary tactics, now prodded their bullocks with the sharp-pointed sticks, and drove at full trot straight towards the combatants. In this manner we gained a position within half a minute that we should perhaps never have obtained had the bucks remained in peaceful tempers; the females perceived the danger of our approach, and they started off, leaping in their usual manner many feet in the air perpendicularly at every bound, leaving the two stupid males in the ecstasy of a mortal struggle.

We reached a position within about 120 yards before the two fools observed us. They at once left off fighting, and having regarded us in astonishment for half a second, one dashed off to the left, and the other to the right, across the open plain devoid of bush, or ruts, or any obstacle to the highest speed.

At that same moment a cheetah that had been held in readiness leapt airily to the ground, and the chase commenced after the right-hand buck, which had a start of about 110 yards. The keeper simply begged us not to follow until he should give the word.

It was a magnificent sight to see the extraordinary speed of both the pursued and the pursuer. The buck flew like a bird along the level surface, followed by the cheetah, who was laying out at full stretch, with its long, thick tail brandishing in the air. They had run about 200 yards, when the keeper gave the word, and away we went as hard as the horses could go over this first-class ground, where no danger of a fall seemed possible. I never saw anything to equal the speed of the buck and cheetah; we were literally nowhere, although we were going as hard as horse-flesh could carry us, but we had a glorious view.

The cheetah was gaining in the course, literally flying along the ground, while the buck was exerting every muscle for life or death in its last race. Presently, after a course of about a quarter of a mile, the buck doubled like a hare, and the cheetah lost ground as it shot ahead, instead of turning quickly, being only about 30 yards in the rear of the buck. Recovering itself, it turned on extra steam, and the race appeared to recommence with increased speed. The cheetah was determined to win, and at this moment the buck made another double, in the hope of shaking off its terrible pursuer; but this time the cheetah ran cunningly, and was aware of the former game; it turned as sharp as the buck; gathering itself together for a final ef-fort, it shot forward like an arrow, picked up the distance that remained between them, and in a cloud of dust for one moment we could distinguish two forms. The next instant the buck was on its back, and the cheetah's fangs were fixed like an iron vice upon its throat.

The course run was about 600 yards, and it was worth a special voyage to India only to see that hunt. The cheetah was panting to an extent that made it difficult to retain its hold. There were a few drops of blood issuing from a prick through the skin of the right haunch, where the cheetah's nails had inflicted a trifling wound when it delivered the usual telling blow of the fore paw, that felled the buck to the ground when going at full speed; beyond this there was no blood, until the keeper cut the throat in the customary manner, and the cheetah, much exhausted, was led to its cage. This was a very exceptional hunt, and a friend who was present declared he had never seen anything to equal it, although he had been all his life in India.

We had several courses, but nothing equalled this exciting hunt. On one occasion the cheetah was slipped at too great a distance, the herd being at least 350 yards ahead. The animal, after a vain effort, was well aware of the impossibility; it accordingly ran up a solitary tree with the agility of a monkey.

From this height the cheetah surveyed the retreating herd of antelopes, and refused to descend when summoned. It was necessary for the attendant to mount the tree, but the difficulty was increased by the cheetah making unamiable faces as the man approached his perch. The wooden ladle was now produced as a lure, and after some hesitation the animal followed the man as he descended; the hood was adjusted over the eyes, and the cheetah was replaced within its cage.

From the description given of the various classes of leopards, the destruction committed by these animals may be easily imagined; fortunately they do not breed like our domestic cats, but they seldom have more than two, or at the most three cubs at a birth. I have always been of opinion that the Government should cease to offer a reward for the destruction of tigers (50 rupees), but that an increased reward should be given for the death of every leopard (25 rupees). The tigers will be always killed by Europeans who do not require the inducement of a bonus, and the sum of 25 rupees would incite the natives to trap and destroy a common pest and scourge (the leopard), which seldom or never affords the hunter a chance of sport.

The cheetah (Felis jubata) should be exempted from this decree, as it seldom attacks domestic animals, but confines its attention to the beasts of the plains and forests.

CHAPTER IX

THE LION (FELIS LEO)

I have left this grand example of the genus Felis to conclude the species, as the tiger is so closely associated with the elephant that I was forced to accord it a place in direct sequence.

In the early days of the world's history the lion occupied a very extensive area; it was common in Mesopotamia, and in Syria, in Persia, and throughout the whole of India. It is now confined to a limited number in Guzerat, and a few in Persia. Beyond these localities it has ceased to exist in Asia. There can be little doubt that, unless specially protected, it will become extinct in Asia within the next hundred years.

Africa is the only portion of the globe where the lion remains lord of the forest, as the king of beasts. The question has frequently been discussed, "Why should the lion have vanished from the scene where in ancient days he reigned in all his glory?" The answer is simple, the lions have been exterminated.

There is a nobility in the character of a lion which differs entirely from the slinking habits of tigers, leopards, and the feline race in general. Although the lion is fond of dense retreats, he exposes himself in many ways, which the tiger seldom or never does, unless compelled by a line of beaters. This exposure, or carelessness of concealment, renders his destruction comparatively easy.

On the other hand, the lioness brings forth a numerous family, generally five or six at a birth, which should keep up the number of the race; in spite of this prolific nature, the lion having from time immemorial been an attraction to the mighty hunter, man has proved too much for him.

The Indian species is considerably smaller than the African variety, and the mane is seldom so dark in colour, or so shaggy. tiger, as the animals differ in form and muscular development. I have never weighed a lion, but I feel convinced that a fine specimen would be heavier than an equally well selected example of a tiger, as the former is immensely massive, especially about the chest and shoulders. The head and neck are larger, although, when boiled and cleaned, the skull does not exceed in size that of an ordinary tiger. It may be safely stated that a lion which measures 9 ft. 8 inches in length would weigh heavier than a tiger of the same dimensions. I have already described that the tiger when springing to the attack does not strike a crushing blow, but merely seizes with its claws. A lion, on the contrary, strikes with terrible strength, at the same time that it fixes its claws upon its victim. The force of this blow is terrific, and many a man has been killed outright as though struck with a sledge-hammer. An instance of this fatal onset deprived me of a most intelligent and excellent German, with whom I was associated during a hunting season in the Soudan.

Florian was a Bavarian who came to Khartoum in the service of the Austrian Mission, employed as a mason. This man had a natural aptitude for mechanical contrivances, and quickly abandoning the Jesuit Mission, after the completion of the extensive convent at the junction of the two Niles, he and a carpenter of the same nation formed a partnership of hunters and traders, establishing themselves at Sofi on the frontier of Abyssinia. They built a couple of circular huts of neatly squared stones, and not only shot hippopotami in the Atbara river, but manufactured extremely good whips from their skins. These were very superior in finish to the ordinary "courbatch" of the Arabs, and they met with a ready sale. Florian excelled as a carpenter, although a mason by profession; he made exquisite camel saddles for the Arab sheiks; these (moghaloufa) were cut from the heart of a tough wood which never warped (Rhamnus Lotus), and were highly prized by the experienced Arabs of the desert. The rainy season was industriously employed in such useful manufactures, and when the dry months arrived, these two excellent men started upon hunting expeditions, and combined business with pleasure.

Although Florian was clever with both head and hands, he was a bad shot; his guns were of a common and dangerous description, one of which burst, and blew his left thumb and forefinger off. After his recovery from this accident he still excelled in work, but he was exceedingly clumsy with his weapons, which were always going off by accident. Upon several occasions these unintentional explosions took place so close to my own head that I suggested it would be safer should he adopt solitary rambles instead of shooting in company.

One night he killed an elephant while watching by moonlight at a drinking-place. On the following morning he sent a trustworthy Tokroori native with an axe to cut out the tusks. The man presently returned with the news that a large lion had eaten a portion of the elephant, and was lying asleep close by, beneath a tree.

Florian immediately gave his man a single-barrelled rifle, and taking a double smooth-bore himself, the two proceeded together towards the spot. Upon arrival at the place where the body of the elephant was lying, the lion was immediately discovered beneath a leafless bush, where it had been seen by the Tokroori. The animal appeared to be thoroughly gorged with elephant's flesh, and, half asleep in the hot sun, it took very little notice of the two men, but remained crouched upon the bare ground, neither grass nor leaves at that dry season existing to form a cover for retreat.

Florian advanced boldly to within about 20 yards, the lion merely regarding him with sleepy astonishment, until he took aim and fired. He missed! The lion instantly assumed an attitude ready for a spring. Florian aimed between the eyes, and again fired. He missed again! The response was immediate: the lion gave a roar, and bounded forward; with a terrific blow upon the head it felled the unfortunate Florian to the ground, and seized him by the neck. Almost at the same moment the faithful Tokroori rushed forward to assist his master, and, afraid to fire lest he should hit him by mistake during the confusion of the struggle, he actually pushed the muzzle of the rifle into the lion's ear and pulled the trigger. The lion fell dead upon the lifeless body of Florian.

Dr. Ori, an Italian in the service of the Egyptian Dr. Ori, an Italian in the service of the Egyptian Government, was at that time purchasing wild animals of the Hamran Arab sword-hunters, and was in camp within a half-hour's march. The Tokroori brought the tragic news, and a party started for the fatal spot. Dr. Ori subsequently described to me the effect of the lion's blow. The skull, which had received its full force, was completely shattered, as if it had been a cocoa-nut struck with a hammer, and several of the lion's claws had penetrated through the bone, as though they had been driven like a nail.

If that had been the attack of a tiger, the skull would not have been injured, although the scalp would have been badly lacerated, and death would have been occasioned by the grip of the jaws upon the neck, not by the blow.

Another instance of the great force of a lion's blow was witnessed by my late friend, Monsieur Lafargue, whom I knew when he was a resident of Berber in the Soudan. This French gentleman was agent to Halim Pasha, the uncle of His Highness Ismail the Ex-Khedive. Halim Pasha was a man of great energy, and he was the first personage in the history of Egypt who sent a steamer from Cairo to ascend the cataracts of the Nile and reach Khartoum. This was accomplished after extreme difficulty in experimenting upon the course of nearly 1600 miles of river, the navigation of which was then unknown to others beyond the native owners of small vessels. Halim Pasha was the first to attempt the commercial development of the White Nile, and Monsieur Lafargue was an admirable representative of his august employer. The steamer arrived safely at Khartoum, and was engaged in the trade of the Blue Nile to Fazocle, and through the White Nile to the unknown, as in those days Khartoum was the southern boundary of Egypt.

Monsieur Lafargue was a charming man, highly educated, with a mind of a peculiar character, that enabled him to lead a happy life in the remote wilderness of the Soudan. It was difficult to understand, when conversing with him in his beautiful house at Berber, or sitting together in his garden on the extreme margin of the Nile, while the desert sands upon the east side of the wall showed the limit of civilisation and fertility, how any man of culture could endure to pass his entire existence in such a narrow boundary—the Nile, the fruitful source, upon one side, and the desert 200 yards beyond; sterile, only because the water could not reach its surface.

He had his books, all the monthly periodicals from Europe, and his newspapers; he also had his private affairs, his agency, which occupied his time; in addition, he had a wife, an Abyssinian lady of great beauty, and of gentle sympathetic disposition. To her husband she was as the moon is to the traveller upon an otherwise dark night. Her story was too romantic and sad to be lightly introduced, but her husband had given up his country, and his family in France, after having made his fortune in the Soudan, entirely upon her account. He described her to me as the "gazelle of the desert, that was contented and happy in its native sands, but would die in the atmosphere of conventional civilisation."

Monsieur Lafargue held a deservedly high position among all classes in the Soudan. He had discovered that no legitimate commerce was possible with the savages of the White Nile; he had therefore advised his employer to that effect, and he had resigned all hope of effecting the original object of his expedition. He was therefore carrying on a business with the native merchants, from whom he purchased gum-arabic from Kordofan, ivory from the White Nile, hides from the Arabs generally, cotton, and cereals, all of which, as opportunity offered, he either sent down the river or across the Korosko desert to Egypt proper.

We were talking about lions, and he told me the following account of what he witnessed as he was returning from the White Nile upon the steamer, then en route towards Khartoum.

The dry season was at its height; all the high grass and other herbage along the river's banks had been burnt by the natives, and the surface of the earth was black and bare. The steamer was going easily down stream, saving her fuel, and as they floated along, with the paddles revolving slowly, a lion was observed upon the dark and lately blackened bank. The vessel was at once stopped, and a trustworthy Tokroori hunter of Lafargue's volunteered to shoot the lion. The man was confident; accordingly he was put ashore, armed only with a single-barrelled rifle.

From the poop-deck of the steamer the whole affair was distinctly visible. They saw the bold Tokroori advance unconcernedly towards the lion, which, although standing when first observed, now immediately crouched. The Tokroori advanced until he was only a few yards distant: he then halted, and fired. With a loud roar the lion flew to the attack, and with a terrific blow it struck the hunter upon the shoulder. The effect was awful; the man was dashed violently upon the ground, and the lion fell across his body; after a few gasps it rolled over and died. The Tokroori never moved.

The steamer was now run alongside the bank, and Monsieur Lafargue, with a number of men, quickly went ashore. Both the Tokroori and the lion were quite dead. The bullet had struck the animal in the chest, and had passed through the heart. The Tokroori's arm was hanging from the hip! It had not only been completely dislocated at the shoulder by the blow, but it had been torn or struck downwards with such extreme force that the flesh had been entirely stripped off the ribs and the side; the arm at the extremity of this ruin was dangling upon the ground, hanging only to the hip by the flesh attached. The Tokroori had been killed on the spot by the shock to the system. This was a remarkable example of force. On the other hand, although the lion frequently uses this dreadful power of striking when in full charge, there are many cases when the animal seizes simply with teeth and claws, like a tiger or others of the race. (A tiger possesses the power to deliver a tremendous blow, but it seldom exercises this force.)

I am of opinion that the act of striking would depend upon the position of the animal or person attacked. There can be no doubt that a lion could fell an ordinary bullock by a blow upon the neck, should it attack from one side, but it would be extremely unlikely that it would strike any horned animal upon the head, as it would risk serious damage to the paw. We have seen that the cheetah strikes the haunch of a black-buck when coursing at full speed, and it is highly probable that the lion would exert its prodigious strength in the same manner, to stun the hind-quarters by the stroke, and, by throwing the animal upon one side, to expose the throat to the grip of the powerful jaws. All beasts of prey occasionally meet with dangerous antagonists, and should the first spring fail, the lion may find an adversary worthy of its fangs in a staunch old African buffalo, in which case the battle would be worth a journey to be witnessed. I once discovered the dislocated skeleton of a buffalo almost intermingled with the broken bones of a lion, the skull of which was lying near, while the skull of the buffalo, devoid of the nasal bones, was lying within a few feet distant, gnawed by jackals and hyenas. The ground had been deeply trampled, showing the desperate character of the recent struggle, which had terminated in the death of both combatants. It is highly probable that two lions had simultaneously attacked the buffalo, who had succumbed after having vanquished one assailant. This is a very common practice among lions, to hunt in company. Mr. Oswell in South Africa had a peculiar example of this when in a day's hunting his friend Major Vardon had wounded a bull buffalo, which had retreated within the forest. The two hunters carefully followed the blood-track, but after a short advance they were startled by a succession of loud roars, which betokened lions close at hand.

There could be little doubt that the wounded buffalo had been attacked; therefore, with proper precaution, they warily approached the spot, until the exciting scene presented itself suddenly on the other side of a large fallen tree, which happily concealed the approach of the two companions.

Three lions were engaged in a life-and-death combat with the gallant old bull, who made a desperate defence, first knocking over one of his enemies, then boring another to the ground, and exhibiting a strength which appeared sufficient to defeat the combination. Suddenly the buffalo fell dead; this was the result of the original wound, as the rifle bullet had passed through the lungs.

The lions were not aware of this, and a quarrel among themselves commenced after their imagined victory. One huge beast reared to half its full height and placed its fore paws upon the body of the prostrate buffalo, while at the head and the hindquarters an angry lion clutched the dead body in its spreading paws, and growled at the possessor of the centre. This formed a grand picture within only a few yards' distance, but a couple of shots from either rifle stretched two lions rolling upon the ground, and the third, terrified at the unexpected reports, bounded into the thick covert and disappeared.

A very good sportsman named Johann Schmidt, a Bavarian who died in my service when in Africa, killed two lions in the act of attacking a giraffe. I saw the skeletons of these animals in the bed of the river Royan a few days after the incident. At that dry season of the year the Royan was devoid of water, except at certain bends where the current had scooped out a deep hole beneath the bank. Johann Schmidt was a poor man, who could not afford the luxury of first-rate rifles; he therefore did his best with most inferior arms, one of which was a light double-barrelled smooth-bore muzzle-loader No.

16. This was a French gun, for which he had given 50 francs at Cairo. By some chance, this common little weapon shot remarkably well with ball and 3 drams of powder. It became his favourite companion. He was strolling one day along the bank of the Royan in Abyssinia, looking carefully down its sandy bed, when he came near to a water-hole in the long intervals, and he suddenly heard the peculiar sounds of a great encounter. The dust was flying high in the air, and as he approached the spot, within the yellow surface of the river's bed, he saw a cloud of sand, in the centre of which was the large body and long neck of a bull giraffe struggling against the attack of two lions. One of these was fastened upon its throat, while the other was mounted upon its hind-quarters, where it was holding on with teeth and claws. Johann concealed himself behind a large tree which grew upon the bank; this abrupt margin was about 20 feet above the river's bed, and not 50 yards from the scene of a hopeless conflict.

The giraffe had no chance; and after a sharp struggle before the eyes of the well-concealed spectator, it was pulled down, and both lions commenced to growl over their contested prey. The position upon a perpendicular bank being thoroughly secure, Johann took a steady shot, and rolled one lion over, close to the dying giraffe; the other looked round for a moment, and sprang up the bank upon the opposite side of the river, but this, being perpendicular, was too high to permit of a direct retreat; a bullet from the remaining barrel struck it through the back, and paralysed the hind-quarters. The animal fell backwards upon the sandy surface of the river, and rolled over helplessly, as the hind legs had lost all power. This gave Johann time to reload, and, seeing that the lion was completely at his mercy, he descended into the river's bed and put a bullet through its head.

The giraffe was still alive, therefore another ball was necessary to complete its despatch; and Johann remained in triumph, having bagged two lions and a giraffe with a gun worth only 50 francs.

I have heard so many tales of lions which have carried away oxen from a kraal, that I have endeavoured to unravel what appears to be a mysterious impossibility. An experienced friend of mine was present when, during the night, a lion bounded over the fence of thorns which formed a protection to the camp, and seizing a full-grown bullock, it jumped the fence, carrying the victim with it.

In the confusion of a night attack the scare is stupendous, and no person would be able to declare that he actually saw the lion jump the fence with the bullock in its grip. It might appear to do this, but the ox would struggle violently, and in this struggle it would most probably burst through the fence, and subsequently be dragged away by the lion, in a similar manner to the custom already described of tigers. It is quite a mistake to suppose that a lion can carry a full-grown ox; it will partially lift the fore-quarters, and drag the carcase along the ground.

Upon one occasion I was strolling through the forest on the margin of the Settite river in Abyssinia, and I suddenly met a large bull buffalo which was exactly facing me, having probably obtained my wind beforehand. It was not more than 20 yards distant, and it threw up its wicked head with the nose pointed directly at me, in the well-known fashion which makes a shot at the forehead utterly impossible. Knowing that my double-barrelled No. 10 with 7 drams of powder would have sufficient penetration, I aimed exactly at the nostril, then fully dilated by the excitement of the animal, and fired. The shot was instantly fatal, as the hard bullet of quicksilver and lead not only passed through the brain, having entered at the nose, but it penetrated far into the neck and cavity of the chest. This was a very large beast, and knowing that the dense covert of nabbuk (Rhamnus Lotus) close by was a great resort of lions, I determined to leave the carcase for the night in the spot where it was then lying.

On the following morning I revisited the place with two of my excellent Tokrooris; we found many fresh footprints of lions in the sandy soil, and a broad trace about 4 feet wide, where the body had been dragged away. This had apparently been effected by more than one lion, as the footprints varied in size.

There was a vast mass of dense green nabbuk growing parallel with the banks of the river. This was an opaque screen of thorny foliage, covering an area of about 200 yards in width, but extending for a great distance. The nabbuk tree bears a small apple the size of a nutmeg, rather sweet, and pleasant to the taste; but the tangled mass, when growing upon the sandy loam near water, is absolutely impenetrable to a human being. Into this secure retreat the lions had crept, forming dark tunnels about 3 1/2 or 4 feet high, for some unknown distance.

The trace of the dragged buffalo led direct to the entrance of one of these obscure tunnels, and there could be no doubt that the carcase was within, and the lions not far distant. I have frequently looked back to absurdities that have been scathelessly committed; among these on more than one occasion I have foolishly ventured upon the exploration of a lion's retreat. With two of my Tokrooris following with spare rifles (all muzzle-loaders) I crept upon hands and knees into the dark tunnel, upon the trace of the dragged buffalo. A light double-barrelled '577 was my companion.

After a few yards the tunnel became much narrowed, and was hardly more than 3 feet 6 inches in height. The bush (evergreen) was so dense that it was very dark, and I could not see any tracks of lions upon the ground over which I crept; cautiously, advancing, with both barrels upon full cock. About 70 yards had been passed in this manner when I distinctly smelt the heavy odour of raw flesh and offal. I looked behind me, and my two men were keeping well together. There could be no doubt that the carcase of the buffalo was not far off, and it was highly probable that the lions would be in forcible possession. We crept forward with extreme caution. The faint and disagreeable smell increased, and was almost insupportable. I presently heard the cracking of a bone, and there could be no doubt that the lions were close at hand. I once more looked round to see if my men were coming on; they were both close up. We crept noiselessly forward for a few yards, and suddenly a dark object appeared to block the tunnel; in another moment I distinguished the grand head and dark mane of a noble lion on the other side of a mass which proved to be the remains of the bull buffalo; another head, of a lioness, arose upon the right, and at the same instant, with a tremendous roar, the scene changed before I had time to fire. We were alone with the remains of the buffalo, and I believe three lions had decamped, never to be seen again in the obscurity of the dense green nabbuk. We were actually in possession, having driven the lions from their prey, simply by our cautious advance, without a shot.

It required some time and trouble to cut off the head of that bull buffalo in the narrow limits of the lion's den, but it hangs upon my walls now as a trophy that might be won from a lion, but never could have been wrested in the same manner from a tiger.

Upon another occasion I crept in a similar manner into one of their dark tunnels, and shot the lion within a distance of four paces, but I never recovered the body, as the animal bounded into the dense thorny substance, which it was impossible for any human being to penetrate. The Hamran Arabs persuaded me to discontinue this kind of exploration, and my Tokrooris having taken the same view of the performance, I gave up the practice, as I did not succeed in actually bagging a lion by the attempt.

In the locality which I have mentioned, the lions, although numerous, were never regarded as dangerous unless attacked; there was an abundance of game, therefore the carnivora were plentifully supplied, and a large area of country being entirely uninhabited, the lions were unaccustomed to the sight of human beings, and held them in respect. During the night we took the precaution to light extensive bonfires within our camp, which was well protected by a circular fence of impenetrable thorns, but we were never threatened by wild animals except upon one occasion.

I was strolling in search of food, with a particular two-grooved single rifle No. 14 which was extremely accurate. Having shot a nellut (A. Strepsiceros), the animal was fixed upon a camel and immediately forwarded to camp, towards which I advanced by a circuitous direction in the expectation of finding other game. The country was perfectly flat in the vicinity of the river, and although much covered with dense bush, it was interspersed with numerous small glades, covered with parched herbage 2 or 3 feet in height. A few Tokrooris accompanied me with spare rifles (all muzzle-loaders, as the breech action had not been introduced in those days), and I was leading the way, occasionally breaking through the intervening bush, with as little noise as possible.

Suddenly, as I was only half emerged from a line of dark green nabbuk, I was surprised by a short roar close to me, and I immediately saw the shoulders and the hinder portion of a lion, the head being concealed by the bush, from which I had not completely emerged. I could have touched it by stretching out my rifle, but personally I was quite unobserved. There was not a moment to lose, and I fired through the centre of the shoulder. With a short roar the lion disappeared; there was a rushing sound in the bushes, and almost immediately another lion occupied the exact position that had been quitted by the lioness. They must have been lying down together when startled by our appearance, or rather by the noise of our approach. This was a splendid chance, but I was unloaded; I stretched my right arm behind me, expecting to receive a spare rifle from my faithful Tokrooris, but they had retreated from the scene, and I remained within 6 feet of a lion's flank with an unloaded rifle and no companion. The lion's head and neck were quite concealed by the dense green bush, and I had no other course to pursue than to reload my rifle. The first tap that I gave the bullet when ramming it home, scared the lion, and with a loud roar it sprang forward and disappeared. My recreant followers now returned, and having administered a few kicks, I took a double-barrelled rifle and we commenced a strict search for the wounded animal. Directed by a low moan, we found her within a few yards, dying; it was a lioness, but there was no trace of her companion, which had been so lately within my reach.

The spare camel was now brought up, and with great difficulty my three Tokrooris, the Hamran Arab, and myself succeeded in placing the lioness across the saddle, having first opened and cleaned the body to reduce the weight.

Blood trickled from the carcase, and dropped upon the ground, thus forming a trace throughout the route until we reached the camp. The lioness was 9 feet 1 inch in length, and, when skinned, the body was dragged to a considerable distance and left for the hyenas.

The fires were blazing after sunset; the horses of my Hamran hunters, and my own, were picqueted within the centre of our enclosure, near the tent, and we were about to retire for the night, when a deep guttural sigh was heard close to the high and impervious fence of kittur thorns. This had been carefully constructed, as life was most uncertain within that questionable district, where the Arab hunting parties invariably killed all natives of the crafty Base tribe whenever met, and they incurred a similar retaliation. The fence was made of entire trees cut off near the roots, and then dragged by the stems into line, with their wide-spreading heads of sharp hooked thorns forming the outside surface; these were locked together by their hooks, entangled, and nothing could possibly have broken through, except an elephant or rhinoceros.

Prowling around this excellent protection was a lion, who was pronounced by my hunters to be the mate of the lioness which I had killed; it was declared that the disconsolate husband had followed the course of his wife's body, denoted by the drops of blood that had dripped upon the ground when carried by the camel towards the camp. My people were of opinion that the lion was determined upon vengeance, and that he would assuredly bound over our fence, although he could not absolutely break through it.

The night was always interesting upon the banks of the Settite river, as vast numbers of wild animals were astir half an hour after sunset, which either came down to drink, or to wander in search of green pasturage, that was only to be found in places from which the water had retreated. The lions were accordingly on the alert, and the threatening sound of their deep voices was to be heard in every direction, until approaching daylight drove them to their thickets.

There is nothing so beautiful, or enjoyable to my ears, as the roar of a lion upon a still night, when everything is calm, and no sound disturbs the solitude except the awe-inspiring notes, like the rumble of distant thunder, as they die away into the deepest bass. The first few notes somewhat resemble the bellow of a bull; these are repeated in slow succession four or five times, after which the voice is sunk into a lower key, and a number of quick short roars are at length followed by rapid coughing notes, so deep and powerful that they seem to vibrate through the earth.

Our nocturnal visitor did not indulge in the usual solo, but he continued throughout the night to patrol the circuit of the camp, occasionally betraying his presence by a guttural roar, or by the well- known deep sigh which exhibited the capacity of his lungs. We could not see to shoot, owing to the darkness outside the fence, and the brightness of our fire within the camp; this my men industriously replenished with wood, and occasionally hurled fire-brands in the direction of the intruder.

At length we went to sleep, leaving the natives to keep watch; they declared that nothing would induce them to close their eyes, as the lion would assuredly carry off one of the party before the morning. To their great discontent, I refused to disturb the night by firing a gun, as I had determined to hunt up the lion on the following day at sunrise.

Upon waking early, we discovered the deep footprints upon the sandy soil, which had marked a well-beaten path around our impenetrable fence, showing that the lion had been patrolling steadily throughout the night. This fact led me to suppose that I should most probably find him somewhere within a very short distance of the camp. I started with some of my best men, and instead of a light single-barrel I carried my '577 rifle.

The position of our camp was exceedingly favourable for game, as the river made a circuitous bend, which had in ages past thrown up a mass of alluvial soil of several hundred acres, all of which was now covered with a succession of dense patches of nabbuk jungle, interspersed with forest trees and numerous small glades of fine dwarf grass, which formed a sward. I felt certain that our visitor of the last night must be somewhere in this neighbourhood, and I determined to devote the entire day to a rigorous search; in this my men were unanimous, as they objected to passing another night in sleepless excitement and anxiety.

Luck was against us. I had numerous opportunities during the day of shooting other animals, but I was devoted entirely to the lion, which we could not find.

I was scratched with countless thorns, as we broke through the thickest bushes, peering beneath their dark shade, and searching every acre of the ground in vain. In spite of the great heat, we worked from early morning until half an hour before sunset without resting from our work; all to no purpose; there were tracks of lions in all directions, but the animal itself was invisible. It was time to turn towards home, and I led the way through low bush and sandy glades not larger than an ordinary room, all of which were so much alike that it was difficult to decide whether we had examined them before, during the day's hard march. In several places we discovered our own footprints, and thus cheerlessly we sauntered homewards, tired, and somewhat disgusted at the failure.

We were within half a mile of the camp, and I was pushing my way through some dwarf green nabbuk about 5 feet high, when, upon breaking into a small open glade, a large lion with a dark shaggy mane started to its feet from the spot where it had been lying, probably half asleep. I instantly fired, before it had time to bound into the thick jungle, and with tremendous roars it rolled over beneath the dense nabbuk bushes, where at this late hour the shade was almost dark. As quick as possible I fired a second shot, as it was rolling over and over, with extraordinary struggles, and it disappeared in the almost impervious bush, dragging its hind legs in such a manner that I felt sure the spine was broken by the bullet. It was so dark that we could not discern the figure of the animal beneath the thorns, although it was only a few feet distant. Having reloaded, I hardly knew what course to pursue; we had no means of driving the lion from the bush, I therefore examined the ground, and we discovered that the nabbuk into which it had retreated was simply an isolated clump, surrounded by narrow glades of sandy turf. From this asylum I felt sure it could not move, and although it would have been more heroic to have crept into the dark cover and have given it a quietus, or more probably to have received it myself, we came to the wise conclusion that if the lion could not move, it would be there on the following morning, when we should have daylight in our favour.

We returned to camp, and the night passed without disturbance. Directly after sunrise we returned to the spot, and we found the lion still alive, although completely paralysed in the hinder portions. A shot in the centre of the forehead terminated the affair, and the joint efforts of ten men succeeded after great exertion in sliding the carcase upon three inclined poles from the ground to the saddle, while the camel was kneeling in a slight hollow, which the people had scraped away for the purpose.

I had no means of weighing this animal, but it was immensely massive, and would according to my estimation have exceeded 500 lbs.

The accounts published respecting the character of lions differ to such a degree that incidents which are considered natural in one portion of Africa may be regarded as incredible in other districts; there can be little doubt that the character of the animal is influenced by the conditions of its surroundings, which renders it extremely difficult to write a comprehensive account, that will embrace the entire family of lions throughout the world. Roualeyn Gordon Cumming gave a terrible description of a night attack upon his camp, when a lion bounded over the thorn fence, and seizing a sleeping servant from beneath his blanket close to the camp fire, carried him off into the surrounding darkness, and deliberately devoured every portion, excepting one leg, which was found on the following morning, bitten off at the knee-joint. This was the more extraordinary, as another man was at the same time asleep under the blanket with the unfortunate victim; this courageous fellow snatched a heavy firebrand from the pile, and beat the lion on the head in the endeavour to save his friend. Instead of relinquishing its prey, the lion dragged the man only a short distance, and commenced its meal so immediately that the cracking of bones could be heard throughout the night.

In southern Africa a night attack by lions upon the oxen belonging to the waggons is by no means uncommon, in books published concerning expeditions to that country, but in nine years' experience of camp life in Africa, both equatorial and to 14 degrees north of the equator, I have never even heard of any actual depredation committed by lions upon a camp or upon a night's bivouac; the nearest approach was the threatening nocturnal visit already described, where no actual damage was inflicted.

There is an instinct natural to all animals which gives them due warning whether man approaches them with hostile intent, and there can be no doubt that every wild animal possesses this discriminating power, and would be influenced according to circumstances. My own experience has led me to an opinion that the lion is not so dangerous as the tiger, although, if wounded and followed up, there cannot be a more formidable antagonist.

Upon several occasions I have seen lions close to me when I have had no opportunity of shooting, and they have invariably passed on without the slightest signs of angry feeling. I was riding along a very desolate path, and a lioness, followed by five nearly full-grown young ones, walked quietly from the jungle, and they crossed within a few yards of my horse's head, apparently without fear or evil disposition. I well remember, at the close of a long march we halted beneath a large tree, which I considered would form an agreeable shade for our tent. I gave my rifle to a servant, who deposited it against the tree, preparatory to my dismounting, when a lioness emerged from the bushes, and walked unconcernedly through our party, within only a few feet of the startled horses. She disappeared without having condescended to increase her pace.

Upon another occasion I had fired the grass, which had left a perfectly clean surface after the blaze. The night was bright moonlight, and I was standing in front of the tent door, when a large, maned lion and a lioness crossed the open space within 10 or 12 yards of my position, and stood for a few moments regarding the white tent; they passed slowly forward, but had disappeared before I had time to return with a rifle.

I once saw a wounded lion decline a challenge from a single hunter. It is possible that a tiger might have behaved in the same manner, but it would be dangerous to allow the opportunity. I had taken a stroll in the hope of obtaining a shot at large antelopes, to procure flesh for camp, and I was attended by only one Arab, a Hamran hunter armed with his customary sword and shield. Having a peculiar confidence in the accuracy of a two-grooved single rifle of small bore, I took no other, and we walked cautiously through the jungle, expecting to meet some animal that would supply the necessary food. We had not walked half a mile when we emerged upon a narrow glade about 80 yards in length, surrounded by thick bush. At one end of this secluded and shady spot an immense lion was lying asleep upon the ground, about 70 yards distant, on the verge of the dense nabbuk.

He rose majestically as we disturbed him by our noise in breaking through the bushes, and before he had time to arrange his ideas, I fired, hitting him through the shoulder. With the usual roars he rolled several times in apparent convulsive struggles, until half hidden beneath the dense jungle; there he remained.

If I had had a double rifle I could have repeated the shot, but in those days of muzzle-loaders I had to reload a single rifle, and as usual, when in a hurry, the bullet stuck in the barrel and I could not drive it home.

In this perplexity, to my astonishment my Arab hunter advanced towards the wounded lion, with his drawn sword grasped firmly in his right hand, while his left held his projected shield, and thus unsupported and alone, this determined fellow marched slowly forward until within a few yards of the lion, which, instead of rushing to attack, crept like a coward into impenetrable thorns, and was seen no more. The Arab subsequently explained that he had acted in this manner, hoping that the lion would have crouched preparatory to a spring; he would then have halted, and the delay would have given me time to load.

I have before remarked upon the extreme danger of despising an adversary, and although I do not consider the lion to be so formidable or ferocious as the tiger, that is no reason for despising an animal which has always been respected from remote antiquity to the present day. It is impossible to be too careful when in pursuit of dangerous game. My friend Colonel Knox of the Scots Fusilier Guards, an experienced and fearless sportsman, very nearly lost his life in an encounter with a lioness, although under the circumstances he could hardly be blamed for want of due precaution. He had shot the animal, which was lying stretched out, as though dead. Being alone, he returned to camp to procure the necessary people, and together with these he went to the spot where he found the lioness in the same position. Naturally he considered that it was dead, but upon approaching the prostrate body he was instantly attacked, knocked down, and seized by the back; he would assuredly have been killed had he not been assisted by his followers. Although he killed the lioness, he was seriously mauled, and was laid up for a considerable period in consequence.

It would be easy to produce cases where lions have caused terrible fatalities, and others where they have failed to support their reputation for nobility and valour; but as I have already observed, there is no absolute certainty or undeviating rule in the behaviour of any animal. The natives of Central Africa, who are first-rate sportsmen, have no fear of the lion when undisturbed by hunters, but they hold him in the highest respect when he becomes the object of the chase. I have known a lion which, when stopped by the nets in one of the great African hunts, knocked over five men, all of whom were seriously wounded, and, although it was impaled by spears, it succeeded in evading a crowd of its pursuers.

Stories of lions are endless, and were they compiled, a most interesting work might result, but my object in producing a few anecdotes, mostly of my own personal experience, is to elucidate the character of the animals by various examples, which prove the impossibility of laying down any fixed or invariable rule.

There can be no doubt that the mode of hunting generally adopted in Central Africa is far more dangerous than the careful contrivances of India, where the tiger, as fully described, is hunted either upon elephants or by posting the guns in secure positions. Even in Rajpootana, where hunting is frequently conducted upon foot, the ground is specially favourable among deep and precipitous ravines, where abrupt rocks and perpendicular banks afford protection to the hunter.

In Central Africa the climate and fodder are so detrimental to horses that the explorer quickly discovers the utility of his own legs, and no experience is so conducive to steady and accurate shooting as the knowledge of an impossibility to escape by speed. We are all creatures of habit, and are more or less the slaves of custom; this is proved AD ABSURDUM by the peculiar feeling when a man who is accustomed to shoot tigers from the secure and lofty position in a tree, finds himself compelled to seek the animal upon foot. In Africa, also in Ceylon, the hunter is so much in the habit of standing upon his own legs that he ceases to fear the attack of any creature, feeling certain of the accuracy of his rifle; but this same individual would begin to feel unnaturally exposed if, after a continuous experience in secure mucharns and mounted upon elephants, he should be suddenly called upon to seek a wounded tiger or lion upon foot. I have never followed lions except on foot. They are killed by the Hamran Arabs on horseback; fairly hunted by two or three of these splendid fellows, and cut down by a stroke across the spine with the heavy broadsword.

The lion is never specially sought for by the natives of Central Africa, but should he be met with in their ordinary hunting expeditions, he takes his chance like all other animals, and is attacked either with arrows or the spear.

Many of the natives are exceedingly courageous, and will advance to the attack upon a lion with spear and shield, or even without the latter safeguard, as they are confident in the support of their companions in case of an emergency. I remember upon one occasion I had wounded a lioness by a shot in the chest from a very accurate but extremely ineffective rifle, which, although '577, carried a small charge of 2 1/2 drams of powder. The animal took refuge in a patch of high grass only a few yards square. Invisible in this retreat, my three hardy natives offered to go in and throw their spears at her, provided I would be ready to support them should she charge into the open when they had failed. This proceeding would have been a reflection upon our superior weapons, and I declined the proposal, as too dangerous to the men. I sent the natives to the summit of a white ant-hill about 7 feet high; from this they espied the animal lying in the yellow grass, but so indistinct that it was impossible to determine her exact position. I accordingly instructed the men to keep a sharp look-out, and to throw their spears should the lioness charge, as I would provoke an attack by firing a shot at hazard into the long grass. Placing Lieut. Baker, R. N., upon my right, with instructions to enfilade the expected attack, I advanced to within 20 yards of the grass, and fired into the spot she was supposed to occupy. The effect was instantaneous. At the report of the rifle the lioness uttered a loud roar and charged directly upon myself, the most prominent antagonist. I fired the left-hand barrel at her chest, but this miserable weapon had no penetration (it was the first and last that I ever possessed with a hollow bullet); the natives hurled their spears, but missed the flying mark; Lieut. Baker fired right and left with a No. 70 small-bore, which hit, but without effect. Everybody turned and ran at their best speed, as the lioness in hot pursuit was within a few feet of us. A native servant of Lieut. Baker passed me with his master's spare gun in his hand. To snatch this from the man, and to turn round and face the still roaring pursuer, was the work of an instant, and I fired into her chest a No. 12 spherical ball with 4 1/2 drams of powder from an ordinary smooth-bore. To my delight, this rolled her over and checked her onset; but she immediately sprang back to her asylum of yellow grass. We were now reduced to our original position, but I knew the wound would be quickly fatal.

The natives recovered their spears, while we all reloaded, and presently one of our people from the summit of the ant-hill excitedly pointed to an object in the high grass; within a distance of about eight yards I distinguished the back of the head and neck of the lioness. She was looking in the opposite direction; this gave me a fatal opportunity, and a shot in the nape of the neck settled the affair, after a well-contested struggle.

It was impossible to carry this animal, we therefore skinned it, and upon opening the stomach we found the sections of a fawn antelope; these when placed in position showed the entire animal, which she must have eaten a few hours previously. This was so fresh that my natives immediately made a fire and roasted the meat, which they ate with great enjoyment as a feast of victory. (We measured this lioness carefully with a piece of string; she was 9 feet 6 inches from nose to tip of tail.)

I shall say no more concerning lions, but I shall always admire the calm dignity of appearance, the massive strength, the quiet determination of expression, and the NOLI ME TANGERE decision, that represent the character of the nation which has selected this noble animal for its emblem.

I do not venture upon the extensive variety of smaller species of the genus Felis; but there is one in India which I have only observed upon two occasions; this is the colour of a puma, rather long in the leg, with pointed tufts of black hair at the tips of the ears, giving it the appearance of a lynx. I have a skin in my possession which I shot in the Central Provinces of India in 1888. The whole of the genus Felis, from the lion to the ordinary cat, have the same number of teeth-six cutting teeth, six front teeth, and two incisors in either jaw. The tongues are invariably rough, and in the lion and the tiger they are prickly to such a degree that flesh could be licked clean off the bone without the preliminary and impatient process of tearing by the teeth.

The often-questioned thorn in the extreme end of a lion's tail is by no means a fallacy; this is a distinct termination in a sharp horny point, which, although only a quarter of an inch or less in length, is most decided. I do not consider that there is any special use for this termination, any more than there would be for the tuft of black hair which forms the extremity, and which conceals the thorny substance.

CHAPTER X

THE BEAR (URSUS)

This is one of the oldest animals in history, and it has survived the attacks of man far more successfully than the more noble beast the lion. This survival may probably result from the secluded habits of the bear, which cannot be classed among the destroyers, such as the carnivora, although it is dangerous when hunted, and not unfrequently it attacks man without any provocation.

The nature of most animals may be judged by the formation of their teeth; those of the bear declare its omnivorous propensities—

In the upper jaw 12 molars, 2 canine, 6 incisors.

In the lower jaw 14 molars, 2 canine, 6 incisors.

There are so many varieties of the bear that it is impossible exactly to define the food of the species. We see the polar bear (Ursus maritimus), which, living upon seals and fish, differs from all others; the grizzly bear (Ursus ferox) of Western America, which will eat flesh when it can obtain it, but is a feeder upon roots and berries. Nearly all bears are inclined to vegetable food and insects, accepting flesh when they find the freshly killed body of an animal, but not seeking live creatures to kill and eat. The sloth bear of India is an exception to this rule, as it refuses flesh, and lives simply upon fruits, berries, leaves of certain trees, roots, and insects of all kinds, the favourite bonne bouche being the nest of white ants (Termites), for which it will dig a large hole in the hardest soil to a depth of 2 or 3 feet. The molars of bears have a close resemblance to those of a human being, exhibiting a grinding surface for the mastication of all manner of substances. The nose is used as a snout, for turning over stones which lie upon the surface, in search of insects, slugs, worms, and other creatures, as nothing comes amiss to the appetite of a bear.

The claws of the fore paws are three or four inches in length, and are useful implements for digging. It is astonishing to see the result upon soil that would require a pick-axe to excavate a hole. Upon the hard sides of such pits as those made in search of white ants, the claw-marks are deeply imprinted, showing the labour that has been expended for a most trifling prize, as the nest when found would only yield a few mouthfuls. I have never appreciated the name of "sloth bear" given to Ursus labiatus, as it is a creature that works hard for its food throughout the year, and being an inhabitant of the tropics, it never hybernates. This species is very active, and although it refuses flesh, it is one of the most mischievous of its kind, as it will frequently attack man without the slightest reason, but from sheer pugnacity. A full-grown male weighs from 280 to 300 lbs. The skin is exceedingly thick and heavy. The hair is long and coarse, with a bunch upon its back of at least 7 inches in length, but there is a total absence of fur, therefore the hide has no commercial value. The chest is marked by a peculiar pattern in whitish brown, resembling a horse-shoe, which is the mark for aim when the animal rears upon its hind legs to attack. There are five claws upon the fore feet, and the same number upon the hinder paws. Although these are not retractile, neither are they so curved or sharp as those of the genus Felis; they inflict terrible wounds upon a human being, and when the head of a man has been in a bear's grip it has generally been completely scalped. I have heard of more than one instance where the scalp has been torn from the back of the neck and pulled over the eyes, as though it had been a wig.

The Ursus labiatus seldom produces more than two or three at a birth, and the young cub is extremely ugly, but immensely powerful in limbs and claws. I have seen a very young animal which held on to the inside of its basket when inverted, and although shaken with great force, nothing would dislodge its tenacious clutch; this specimen was about six weeks old.

Although many varieties of bears are tree-climbers, there are others which are contented with the ground, and which could not ascend a tree even should they be tempted by its fruit. The grizzly bear (Ursus ferox) belongs to this class, and his enormous weight would at any time necessitate especial care when experimenting upon the strength of boughs. I do not believe that any person has actually weighed a grizzly, but an approximate idea may be obtained through a comparison with the polar bear (Ursus maritimus), which is somewhat equal in size, probably superior. When I was in

California, experienced informants assured me that no true grizzly bear was to be found east of the Pacific slope, and that Lord Coke was the only Britisher who had ever killed a real grizzly in California. There are numerous bears of three if not four varieties in the Rocky Mountains, and these are frequently termed grizzlies, as a misnomer; but the true grizzly is far superior in size, although similar in habits, and his weight varies from 1200 to 1400 lbs.

Mr. Lamont, in his interesting work Yachting in the Arctic Seas, gives the most accurate account of all Arctic animals that he killed, and having the advantage of his own yacht, he was able to weigh the various beasts, and thus afford the most valuable information in detail. This is his account of a polar bear (Ursus maritimus) which he himself killed :-

"He was so large and heavy that we had to fix the ice-anchor, and drag him up with block and tackle, as if he had been a walrus. This was an enormous old male bear, and measured upwards of 8 feet in length, almost as much in circumference, and 4 1/2 feet at the shoulder; his fore paws were 34 inches in circumference, and had very long, sharp, and powerful nails; his hair was beautifully thick, long, and white, and hung several inches over his feet. He was in very high condition, and produced nearly 400 lbs. of fat; his skin weighed upwards of 100 lbs., and the entire carcase of the animal cannot have been less than 1600 lbs."

This weight is equivalent to a large-sized English cart-horse. I have seen one of the skins procured by Mr. Lamont, and I can readily appreciate his account of the weight. I have also seen a skin of a grizzly bear killed at Alaska by Sir Thomas Hesketh; this was cured by Mr. Rowland Ward, who showed it to me at his establishment, 160 Piccadilly, and it was very little inferior to the skin of the polar bear. I quite believe the accounts I have received in California are correct, and that the grizzly may sometimes exceed 1400 lbs. in weight. There is a considerable difference in size between the male and female, the former being superior. Like all other animals, the mother is particularly attached to her young, and when in company with them she is more than ordinarily ferocious, as she appears to suspect every stranger of some hostile intentions towards her offspring.

The increase of population in many countries has resulted in the destruction of all animals that were considered dangerous to man; thus the wolf and the bear have both disappeared from Great Britain, and they have become scarce in France.

Thirty-five years ago, I was in a wild portion of the Pyrenees, in the hope of finding bears at the first snows of winter, when by extreme bad luck a fall took place so suddenly and severe that a pass was blocked, which prevented my arrival at a narrow valley, between the lofty mountains named Tram-Saig. I had been assured that the bears would hybernate at the commencement of winter, and that they could only be found at the season when the first snow-fall would expose their tracks.

On the following day I managed to get through the pass, and to my intense disgust, upon arrival, I found that I was a day too late, as the Maire, who was a great chasseur, had killed two bears, a mother and half-grown young one, on the preceding day, thus verifying the information I had received.

I saw the freshly killed skins pegged out to dry, and a few days later I ate a portion of the paws in an excellent stew when dining with the Prefect of Bagneres-de-Bigorre, to whom they were forwarded as an esteemed present.

The larger bear-skin gave me the impression that the original owner must have been the size of a heifer twelve or fifteen months old. This was the ordinary brown bear of Europe, which still exists in Transylvania, Hungary, Italy, and especially in Turkey. The same bear inhabits Asia Minor, and both these varieties hybernate at the commencement of winter. In the extensive forests and mountains about Sabanja, beyond the Gulf of Ismid, I have seen the wild fruit trees severely injured by the brown bears, which ascend in search of cherries, plums, apples, walnuts, and sweet chestnuts. The heavy animal knows full well that the extremity of the boughs will not support its weight, it therefore stands erect upon a strong limb and tears down the smaller fruit-laden branches within its reach. Although bears are numerous throughout the forests, there is only one season when they can be successfully hunted; this is in late autumn, when the fruits are closing their maturity, and the apples and nuts are falling to the ground. The bears then descend from the mountain heights, and may be found late in the evening or before sunrise in the neighbourhood of such food.

Asia Minor and Syria possess two distinct varieties of bears, although the countries are closely connected, and these animals are not inhabitants of the same district. The Syrian bear is smaller than the ordinary brown bear, and would hardly exceed 300 lbs. in weight. The fur is a mixed and disagreeable colour, a dusky gray of somewhat rusty appearance, but blanched in portions as though by age. This species is to be found at the present day upon Mount Horeb, and the natives assured me that, when the grapes are ripe, it is necessary to protect them by watchers armed with guns, to scare the bears during night.

Wild animals which hybernate have a peculiar instinct for selecting hiding-places, which can seldom be discovered; in these they lie, free from all intrusion.

The fruits of late autumn fatten the bear to a maximum condition, and when the harvest is over, and the ground is covered with a dense sheet of snow, it retires to some well-known cave, high among the mountains, in such undisturbed seclusion that it is seldom visited by the foot of man. Within a cave, nestled in ferns or withered leaves and grass, the fatted bruin curls itself to sleep throughout the winter months, and the warmth necessary to its existence is supplied by its own fat, which, being rich in carbon, supports vitality at the expense of exhaustion of supply.

If the fat bear could see itself previous to hybernation in November, and again be introduced to its own photograph upon awakening from its sleep in March, it would be prepared to swear against its own identity. It arises from its winter's nap in wretched condition, having lived entirely upon capital instead of income. Young shoots, and leaves of spring, wild tubers which it scratches from the ground, detected by its keen sense of smell, together with snails, beetles, worms, and everything that creeps upon the earth, now form the bill of fare, until the summer brings forth the welcome fruits that reproduce the condition which the bear had lost through hybernation.

It is impossible to unravel many of the mysteries of Nature, and the cause which prompts the instinct of a winter's sleep will always remain doubtful. I should myself attribute hybernation to the necessity of repose at a period when food was impossible to procure. The body can exist for an incredible length of time, provided that it is capable of undisturbed rest, which appears in a certain degree to take the place of extraneous nutriment. It is well known that every exertion of the muscles is a loss of power, the force of the body being represented by heat. To lift a weight or to move a limb requires a certain expenditure of heat, which means force; this loss of heat and power is recuperated by food; thus in the absence of provisions for the necessary supply, there would be no loss of heat if there is no exertion. Sleep is the resource, as the body is not only at rest, but the brain is also tranquil; there is accordingly a minimum of exhaustion. Human beings have been known to live without food of any kind (excepting water) for a period of forty days, and have then resumed their ordinary course, simply confining themselves to moderate diet for the first few days after their long abstinence. In a time of starvation in Africa I have frequently composed myself to sleep in the absence of my daily food, and I have awoke without any disagreeable craving for a meal. Continued sleep will to a certain extent render the body independent of other nutriment, and I should imagine that the custom of hybernation has been induced by necessity. At a season when the fruits of the earth are exhausted, the ground frozen to a degree that would render scratching for roots impossible, an animal that was dependent upon such productions for its existence must either starve or sleep. The sleep is in itself a first stage of the process of starvation. The creature that can sleep through an existence of four months without food, and lose the whole of its fat during that interval of inaction, has already lost all that supported life during the period of total abstinence—the fat, or carbon. If it were to begin another turn of sleep in its exhausted state, it would be unable to support its existence.

I therefore regard hybernation as the result of the highest physical condition, the animal being thoroughly fat; the food ceases, and the beast, knowing this fact, lays itself down to sleep, and exists upon its own fat, which gradually disappears during the interval of starvation. The bear wakes up in spring with a ragged ill-conditioned skin, instead of the glossy fur with which it nestled into rest; and it finds its coat a few sizes too large, until an industrious search for food shall have restored its figure to its original rotund proportions.

The proof of this necessity for repose during a period of enforced abstinence will be observed in the independence of tropical bears, which do not hybernate, for the best of all reasons, "that there is no winter," therefore they can procure their usual food throughout every season without difficulty or interruption.

The animals of America are all exaggerated specimens of the species, and the grizzly bear, if standing by the side of the ordinary brown bear of Northern Europe, would hardly exhibit any striking difference except in superior size and a slight roughness of colour. I have heard the question frequently discussed when in the Big Horn range of the Rocky Mountains in Wyoming; some of the professional hunters term all bears grizzlies, while others deny the existence of the true grizzly except upon the Pacific slope.

There is no doubt that all the American bears will eat flesh whenever they can obtain it, although they do not pursue animals as objects for food. The usual custom in bear-shooting is to kill a black-tail deer and to leave the body untouched. If this course is pursued throughout the day, three or four deer may have been shot in various localities, and these will lie as baits for the bears.

At daybreak on the following morning the hunter visits his baits, and he will probably find that the bears have been extremely busy during the night in scratching a hole somewhat like a shallow grave or trench, in which they have rolled the carcase; they have then covered it with earth and grass, and in many cases the bears may be discovered either in the act of working, or having completed their labour, they may be lying down asleep half gorged with flesh, and resting upon their own handiwork. In this position it is not difficult to obtain a shot.

When I was in the Big Horn range in 1881 several shooting parties had preceded me on the two previous seasons, and the bears had been worried to such an extent that they were extremely cautious and wary. There was a small party of professional skin hunters who were camped within a mile of my position, consisting of two partners, Big Bill and Bob Stewart. The latter went by the name of Little Bob, in contrast to his enormous companion. Bob was of Scotch extraction; he was about 5 feet 5 inches in height, very slight, and as active as a cat. In his knowledge of every living creature upon the mountains he was perfect; from the smallest insect to the largest beast he was an infallible authority. Bob was a trapper and hunter; he followed the different branches of these pursuits according to the seasons; at one time he would be trapping beavers and red foxes, at another he would be shooting deer for the value of their hides. This cruel and wasteful practice I shall speak of in another portion of this work.

His only weapon was a single-barrelled Sharp's .450 rifle, and he possessed the most lovely mare, beautifully trained for shooting, and not exceeding 14 1/2 hands in height. Little Bob, on his little mare, would have formed a picture. On one occasion I had returned to camp a little after 5.30 P.M., and as the sun sank low, the deep shadows of the hills darkened our side of the narrow glen, and by 6 o'clock we were reduced to a dim twilight. Presently, in this uninhabited region, a figure halted within 15 paces of our tent, which was evidently Bob Stewart, mounted upon some peculiar animal of enormous bulk, but with a very lovely high-bred-looking head. This was Bob's pretty mare, loaded, and most carefully packed with the trophies of his day's sport, as a solitary hunter, quite alone and unaided since 8 A.M. His pony carried the skins of three bears and four black-tail deer, which he had shot, skinned, and packed upon his sturdy little companion.

The bears consisted of a mother and two half-grown young ones of the choice variety known as "silver-tipped." He had come across the family by chance while riding through the forest, and having shot the mother through the shoulder, she fell struggling between her cubs; these pugnacious brutes immediately commenced fighting, and a couple of shots from the rapid breechloading Sharp rifle settled their ill-timed quarrel.

Bob was the most dexterous skinner I ever saw; he would take off a skin from a deer or bear as naturally as most persons would take off their clothes; and the fact of a man, unassisted, flaying seven animals, and arranging them neatly upon the Mexican saddle, would have been a tolerable amount of labour without the difficulty of first finding and then successfully shooting them.

The hide of the largest bear would weigh fully 50 lbs., those of the smaller 25 lbs. each = 100 lbs. The four black-tail deer would weigh fully 50 lbs. Therefore the mare was carrying 150 lbs. of hides, in addition to Bob Stewart, who weighed about 9 stone, making a total of about 276 lbs., irrespective of his rifle and ammunition.

It was a strange country; the elevation of our camp was about 10,000 feet above the sea-level, although we were in a deep and narrow glen, close to a very small stream of beautifully clear water. Upon either side the valley, the hills rose about 1400 feet; at that season (September) the summits were in some places capped with snow. The sides of the hills, sloping towards the glen, were either covered with forests of spruce firs, or broken into patches of prairie grass and sage bush, the latter about as high as the strongest heather, and equally tough and tiresome.

The so-called camp was upon an extremely limited scale; a little sleeping tent only 7 feet by 7, and 5 feet 8 inches in the highest portion; this had no walls, but was simply an incline from the ridge-pole to the ground; it was a single cloth, without lining of any kind, and bitterly cold at night. This was rough work for a lady, especially as our people had no idea of making things comfortable, or of volunteering any service. If ordered to come, they came; to go, they went; to do this or that, they did it; but there was no attempt upon their part to do more than was absolutely required of them. Shooting in the Big Horn range is generally conducted upon this uncomfortable plan. It is most difficult to obtain either men or animals; but, although useless fellows for any assistance in camp, they were excellent for looking after the horses and mules, all of which require strict attention.

We had only four men, all told — my hunter Jem Bourne, the cook Henry (a
German), Texas Bill, who was a splendid young fellow, and Gaylord.

Although I have travelled for very many years through some of the roughest portions of the world, I have always had a considerable following, and I confess to disliking so small a party. Including my wife, we were only six persons, and it was impossible to consume the flesh of the animals killed. I cannot shoot to waste; therefore upon many occasions I declined to take the shots, and thus lost numerous opportunities of collecting splendid heads; this destroyed much of the pleasure which I had anticipated. There were no Indians, as they are confined to their reservations; therefore it was almost criminal to destroy wantonly a number of splendid beasts, which would rot upon the ground and be absolutely wasted. Several parties of Englishmen had not been so merciful; therefore the Americans had no scruples, and commenced an onslaught, general and indiscriminate, shooting all animals, without distinction of age or sex, merely for the value of the skins; the carcases of magnificent fat deer were left to putrefy, or to become the food of the over-satiated bears, which themselves fell victims in their turn.

This was the slaughter in which Bob Stewart and Big Bill were engaged in partnership. They never shot in company, but each started upon his independent course at 8 or 9 o'clock A.M., after having employed themselves since daylight in pegging out the skins to dry, that had been shot on the previous day. The most valuable of the deer-skins was the black-tail, which realised, at a price per lb., 11s. This hide is used for making a very superior quality of glove, much prized in California.

I strolled over to the camp of the two partners one morning, as I was on the way to shoot, and I found them engaged in arranging their vast masses of skins, all of which were neatly folded up, perfectly dry, without any other preparation than exposure to the keen dry air of this high altitude.

Upon my inquiry of Big Bill respecting his operations on the previous day, he replied that he "guessed he had been occupied in running away from the biggest grizzly bear that ever was cubbed."

Big Bill was a Swede by parentage, born in the States. By trade he was a carpenter, but he had of late years taken to skin-hunting. He was an enormous fellow, about 6 feet 3 or 4, with huge shoulders and long muscular arms and hands. There was no harm in Bill; he was a first-rate shot with his .450 Sharp rifle, which appeared to be the weapon in general favour; but he had met with an adventure during the previous year which made him rather suspicious of strangers.

Somewhere, not far from his present camp, a mounted stranger dropped in late one evening. The man was riding a good horse, but was quite alone; so also was Big Bill. The camp of the skin-hunter was then the same in appearance as when I saw him and his partner Bob Stewart—simplicity itself; a long spruce pole was lashed at either end to two spruce firs; against this, leaning at an angle of about 45 degrees, were sixty or seventy straight poles laid close together, and upon these were arranged spruce boughs to form a thatch. This lean-to provided a tolerable shelter within the forest, when the wind was sufficiently considerate to blow at the back against the thatch, instead of direct towards the open face. The ground in the acute angle was strewed with branches of spruce, and a large fire was kept burning during night, exactly in front, the whole arrangement exhibiting the principle of a Dutch oven.

In such a camp, Big Bill received the stranger with the hospitality of the wilderness, and they laid themselves down to rest in the close companionship of newly-made friends.

The morning broke, and as Big Bill rubbed his eyes with mute astonishment, he could not see his friend. He rose from his sleeping-place, and went outside in the cold morning air; he could not see his horses. A horrible suspicion seized upon him; he searched the immediate neighbourhood; the animals had vanished, both horses and mules were gone, together with the unknown stranger, to whom he had given food and shelter for the night.

Fortunately there was a particular horse which Big Bill for special reasons kept separate from the rest; this animal was picqueted by itself among the spruce firs at some little distance, and had been unobserved by the departed stranger. To saddle the horse, and to follow in pursuit at the highest speed upon the trail of the horse-stealer, was the work of only a few minutes. The track was plain enough in the morning dew, where ten or a dozen mules and horses had brushed through the low prairie grass. Big Bill went at a gallop, and he knew that he must quickly overtake them; his only doubt lay in the suspicion that there might be confederates, and that a strong party might have joined together to secure the prize, instead of the solitary stranger being in charge. However, at all hazards he pushed on at best speed in chase; at the same time, the horse-stealer, thoroughly experienced in his profession, was driving his ill-gotten herd before him at a gentle trot, thoroughly convinced that it would be impossible to be overtaken, as the owner had been left (as he supposed) without a horse.

At length, after a pursuit of some hours, upon attaining the summit of a broad eminence, Big Bill's eyes were gladdened by the sight of some distant objects moving upon the horizon, and he at once redoubled his speed.

The stranger, innocent of suspicion, trotted leisurely forward, whistling, and driving his newly acquired animals with professional composure, without condescending to look back, as he felt certain of security, having left his hospitable friend of the preceding night with nothing better than his own legs for locomotion.

In the meantime, Big Bill was coming up at a gallop; he was boiling with indignation at the treacherous conduct of his uninvited guest; and being fully alive to the manners and customs of the West, he placed his Sharp rifle upon full-cock to be in readiness for an explanation.

A few minutes sufficed to shorten the distance to 100 yards, when the astonished horse-stealer was surprised by the sound of hoofs upon the stony soil, and, turning round, he was almost immediately confronted with the threatening figure of Big Bill. The dialogue which ensued has not been historically described; there was none of the bombast that generally preceded the combats of Grecian heroes; but it appears that the horse-stealer's right hand instinctively grasped the handle of his revolver, not unseen by the vigilant eyes of Big Bill, who with praiseworthy decision sent a bullet through his adversary's chest from the already prepared Sharp .450; leaving the lifeless body where it fell, he not only recovered all his stolen animals, but also possessed himself of the horse and saddle which only recently belonged to the prairie horse-stealer without a name.

The gigantic Swede returned to his solitary camp, well satisfied with his morning's work, as he had gained instead of losing, and he had saved the State of Wyoming the expense and trouble of hanging a man for a crime which is supposed to deserve no mercy, that of "horse-stealing."

Of course this instance of determination and extreme vigilance gained for Big Bill the admiration of the extremely limited number of people who would be called "the public" in the outlying portions of Wyoming; but although contented with himself, Big Bill was always suspicious of a solitary stranger, as he had an undefined idea that some relative of the defunct horse-dealer might draw a trigger upon him unawares. It was this redoubtable Big Bill who now confided to me that he had been running away from some monster grizzly bear only on the preceding day. He pointed out the spot, as nearly as possible, from where we stood during his narrative. "There," he said, "do you see that low rocky cliff on the tip top of the hill just above us? That was the place just beneath, on that little terrace-like projection with a few spruce firs upon it. There's a steep but not a difficult way down by the side of that cliff, and when young Edmund and I got down upon that terrace, there were a lot of big rocks lying about, and all of a sudden one of 'em stood up on end within 10 yards of me, and sat up regularly smiling at me, with the most innocent and amiable expression of countenance I ever saw. That was the biggest grizzly bear I ever came across; he was as big as the biggest bull I ever saw in the ranche, and there he was, sitting up on end like a dog, and almost laughing. There was no laugh in me, I can tell you; I just lost no time, but turned round, and hooked it; and I don't think I ever ran so fast in all my life."

"But why did you not shoot him?" I exclaimed with astonishment. "Shoot him? Oh yes, that's very likely, when he wasn't farther than 10 yards off, and I should have had such a poor start, and no place to run to! No, I knew better than that, with a single-barrel Sharp .450. If I had had your double-barrel .577, with a big solid bullet, and 6 drams of powder, I shouldn't have run away; but I go hunting for skins with my little Sharp, and I don't want a grizzly to go hunting for my skin; not if I know it. I've left him for you, and d'ye see, if you go up there this morning, there's some snow about, and you'll likely come across his tracks. If you do, you'll be astonished, I can tell you."

Ten minutes after this discourse, I was on my way up the mountain side in the hope of meeting this extraordinary bear.

Upon arrival at the summit, there was a splendid view of the main range of the Rocky Mountains, about 70 miles distant, across a desolate region some 4000 feet below the point upon which we stood. There was a little snow, but only in patches on the mountain top, and, when near the terrace upon which Big Bill had had his interview with the bear, we certainly discovered an enormous track, the largest that I have ever seen.

We attempted to follow this for some hours, but to no purpose; on several occasions I could have taken deadly shots at black-tail deer and wapiti, but I determined to reserve my bullet for the big game, the object of our pursuit. The day passed away in failure. The next day was equally disappointing; from morning to sunset I fagged over the summits and the spruce fir sides of the mountains, without a trace of the big bear. We passed the old traces that we had seen the previous day upon the snow, but they were still more indistinct, and there was nothing fresh. I was determined, if possible, to find this bear, therefore I devoted a third day to the pursuit, discarding all other game. On the third morning I started with Texas Bill and Jem Bourne, all mounted, and we rode by a circuitous route to the summit of the hill above the valley of our camp. The snow had melted in most places, leaving only small half-thawed patches. We had so thoroughly explored the entire hillside for a distance of several miles during the last two days, that I arranged a beat on the other side of the mountain, upon the northern slope, facing the far-distant Rocky Mountains.

There were no spruce forests upon this side, but the long incline was merely a sheet of rough prairie grass about 18 inches high, intersected by deep ravines, filled with dwarf cotton-wood trees, resembling the silver-barked black poplar. These trees grew about 25 feet high, and as thick as a man's arm, but so close together that it was difficult to force a way through on horseback.

There were many isolated patches of this covert in various places upon the face of this northern slope, all of which were likely to harbour bears or other game. My eye caught instinctively a long dark ravine which cut the mountain from top to base, extending several miles; this was intersected about a mile and a half from the summit by a smaller ravine, also springing from the drainage of the highest ridge, and at the point of junction the two formed a letter Y, the tail continuing, widened by the increased flow of water. There was at this season a very slight stream about an inch in depth, which resulted from the melting of the small amount of snow upon the heights.

There could not be a more likely place for bears, and I instructed my two men to ride to the bottom of the ravine, and to force their horses through the thornless thicket, making no other noise, but occasionally to tap the stems of trees with the handles of their whips.

I dismounted, and my well-trained horse followed close behind me down the steep hillside, exactly on the border of the ravine. This was not more than 80 yards across; thus I could command both sides should a bear break covert, when disturbed by my two beaters; there could not have been a more favourable locality.

My men were thoroughly experienced, and the noise made by the horses in struggling over stones and in rustling through the cottonwood trees was quite sufficient to disturb any animals that might have been there; accordingly they seldom tapped the tree-stems.

Black-tail deer were very plentiful; these were about the size of an ordinary fallow-deer, and they were extremely fat and delicious venison; but their horns were still in velvet, and would not be clean until October. I could have shot several of these animals; but I was full of good resolutions to resist all temptation, and to restrict my shooting to the long-sought bear.

We had followed the course of the ravine for about a mile, when I suddenly heard a tremendous rush among the cotton trees beneath me on the right, followed by excited shouts—"Look out! look out! A bear! a bear!"

I halted immediately, and in a few seconds three splendid wapiti stags broke covert about 100 yards before me, and at full gallop passed across the open ground by which I was descending. My good resolutions crowded upon me as I instinctively aimed at the stag with the finest head, and I resisted the temptation nobly until they were nearly out of sight, passing down a hollow on my left about 150 yards distant. Somehow or other I pulled the trigger; a cloud of dust suddenly arose from the spot where the three stags had disappeared, and I felt sure that the wapiti was down.

At the sound of the shot my men struggled up the steep ascent and joined me. "Why did you shout 'A bear! a bear!'?" I asked. — "It was a bear, wasn't it? I saw a great brown rump for a moment, and I thought it was the bear." — "No bear at all," I answered, "and I have been fool enough to shoot at a wapiti. . . . I think you will find it just in the hollow beneath the ridge."

The men rode to the spot, and sure enough a magnificent stag was lying dead, shot through the shoulder. A wapiti stag weighs about 900 lbs. when fat in August and September. The fat upon the brisket of this animal was 5 inches thick, and that upon the rump and loins was nearly 3 inches. We cut this off in one complete piece, and when cold, within half an hour it stood up like a cuirass. This was one of the finest that I ever saw, and we took the trouble to cut up all the choicest joints, and concealed them in the branches of a species of yew that was growing upon the edge of the ravine. The delay from my folly in taking this shot exceeded an hour, but the head of the stag was a handsome specimen, and we placed it upon a large boulder of rock, to be sent for upon a future occasion.

We again recommenced our search, comforting ourselves with the reflection that "if the bear was in the ravine, the report of the shot would not affect it; and if it was not in the ravine, it would not matter."

As we continued the descent of the mountain slope, the ravine grew wider, and it was now quite 100 yards across; this would increase the probability of finding game, as there was a larger area of covert at the bottom. I was walking carefully in front of my horse, when, without any alarm given by my men from the bottom of the ravine, my attention was attracted by a rushing sound in the dense cotton trees, and I observed several that were in the thickest part shaking in an extraordinary manner, as though an elephant or a rhinoceros was rubbing itself against the stems.

I ran forward towards the spot, and within 15 paces of me I saw a wapiti stag caught by the horns; these were completely entangled among the stems of the thickly growing trees, and the splendid beast was taken prisoner. I could only see occasionally a portion of the horns, and then, as it struggled to escape, I caught sight for a moment of a head and neck sufficient to prove that it was a very splendid beast, with beautiful spreading antlers. The animal was almost within my grasp, and I could have shot it with a pistol; but my good resolutions stood firm. I refused the shot, as we had meat of the finest quality that would keep for a week, and to kill another wapiti would be mere waste of life. In a couple of minutes occupied with this human reflection, yet sorely tempted to take the shot, the stag broke loose, and I heard it crashing full speed down the ravine, and my men shouting loudly that I should "look out!"

Hardly two minutes elapsed before I saw, at about 300 yards' distance, the most magnificent stag that I have ever seen. This splendid beast issued from the ravine, and exhibited a pair of antlers that, large as the animal was, appeared quite disproportioned to its size. They resembled the wintry appearance of a large branch from an oak tree, and this was the prize which I could not distinctly see when entangled in the cotton-wood, within my grasp. This noble stag descended the mountain side at full speed, and I watched it with longing eyes until it was completely out of sight, fully determined that I would never indulge in good resolutions again, that humanity was humbug, philanthropy puerile, and that the rule of success depended upon the principle "Never lose an opportunity."

I was fairly disgusted with myself, and calling my men, I described to them the magnificence of my lost stag. Instead of consolation they said, "Well, if you're come all this way to shoot, and you won't shoot, I don't quite see the use of your coming." That was all I received as a reward for having spared an animal's life which I did not wish to sacrifice wantonly.

"All right; go back and drive the covert to the end; you may depend upon it I'll take the next shot, whatever it may be." The men rode down the steep sides of the ravine, and we recommenced our beat.

Nothing moved for some time, and I mounted my horse as we were approaching the junction of the smaller ravine on my left, which formed the letter Y. I was about 100 yards ahead of my two men, and I descended into the stony depression, crossed the little stream, and ascended the opposite side with some little difficulty, as it was extremely steep, and, together with my 12 lb. rifle, cartridges, and a 26 lb. Mexican saddle, I rode about 18 stone. We reached the top, from which I could look down into the larger ravine on my right, and the lesser on my left, but a number of large rocks, 3 or 4 feet in height, and others of smaller size, made it difficult for my horse to thread his way. Just at this moment I heard the report of a revolver and shouts in high excitement—"The bear! the bear!" Before I had time to dismount in the awkward position among the rocks, I saw a large bear within two yards of me, as he had run at full speed up the steep bank from the bottom of the ravine without having observed me, owing to the rocks; he therefore passed close to my horse upon the other side, only separated from us by the large rock between. In an instant the bear, having seen the horse, turned to the left, and dashed down hill into the smaller ravine which I had just crossed. I jumped off my horse, and ran along the edge, ready to take a shot the moment that I could obtain a clear view of the bear, which I could see indistinctly as it ran along the bottom of the channel, in which was the trickling stream. As I followed, always keeping the animal within view, I felt certain that it would presently forsake this narrow gully, and would cut across the open to regain the large ravine from which it had been dislodged. I therefore raised the 150 yards sight as I ran along the edge, to be in readiness should it try the open. The bear kept me running at my best to keep it in sight, and I was just beginning to think it advisable to fire through the intervening bushes, when, as I had expected, it suddenly turned to the left, ran up the bank with extreme activity, and appeared upon the steep open grass-land, with the intention of cutting across to the larger hiding-place. This was a splendid chance, as the dark colour of the bear looked well upon the yellow grass. I made a most satisfactory shot with the .577 at 150 yards, the bullet passing through the kidneys, and the bear rolled over and over the whole way down the steep grassy hill, until stopped by the thick bushes, which alone prevented it from rolling into the streamlet at the bottom.

My two men came galloping up, and shortly dismounted, and we all descended to the place where the bear was lying, almost dead. In fact, it died while we were standing over it.

"Well done; that was a fine shot, and we've got the grizzly bear at last," exclaimed Jem Bourne. "THE bear? This is not the bear that Big Bill ran from," I replied; "impossible, this is a silver-tip, and not a true grizzly." The argument that ensued over the carcase of that bear was quite enough to make me an unbeliever in the ordinary accounts of native hunters. I calculated that the body weighed about 600 lbs., as my two men were 6 feet high, and exceedingly powerful, and our united efforts could not move the bear one inch from the spot where it had fallen; it may have exceeded that weight, as it was full of fat, and in the finest condition. We skinned it, and had some trouble to induce the horse to permit the hide to be lashed upon its back. Although a fine bear, Big Bill on our return would not acknowledge that it could be compared with the monster which he had seen with such "a smiling countenance." I was quite of his opinion, as the tracks which I saw in the snow were very much larger than the paws of the bear that I have described.

The foot of a bear leaves a print very similar to that of a human being who happens to be flatfooted, but the breadth is larger in proportion to that of a man. It is a curious fact, that a shot through the kidneys of any creature occasions almost instantaneous death, and the animal falls immediately, as though shot through the neck; this proves the terrible shock to the system, as the body is smitten with a total paralysis.

The opinions of professional hunters differ in such an extraordinary manner upon the question of bears, that it would be impossible for a mere visitor to arrive at a satisfactory decision. It is admitted by all that the grizzly bear is the monarch; next to him in size is the cinnamon bear, named from the colour of its fur; No. 3 is the silver-tipped; and No. 4 is the black bear.

The question to be decided remains: "Is the cinnamon bear the grizzly, with some local difference in colour?" My people called the silver-tipped bears "grizzlies," which was an evident absurdity; but, as they were men experienced in the Big Horn range, it was difficult to disbelieve their evidence concerning the occasional presence of a true grizzly. I found, whilst riding through an extensive forest of spruce fir, an enormous skull of a bear, the largest that I have ever seen, except that of the grizzly, compared with which all others were mere babies; what could this have been, unless a true specimen of that variety?

There can be little doubt that bears of different kinds intermingle occasionally by cross breeds, and many are met with which do not exactly correspond with the colouring which distinguishes the varieties already mentioned; but in my opinion those distinct varieties actually exist, and any departure occasioned by cross breeding is simply an accident. Eighteen months before my visit to the Big Horn range, the present Lord Lonsdale, together with a large party, was hunting upon the same ground, and at that time the country, being new to British sportsmen, was undisturbed. The bears were so numerous and unsophisticated that the party bagged thirty-two, and game of all kinds indigenous to the locality was in the superlative. It is astonishing that any game remains after the persistent attacks of gunners, especially in such countries, where open plains expose the animals to the sight of man. In the Big Horn range, at high altitudes of from 8000 to 12,000 feet, the open grass prairie-ground predominates. There are plateaux and hill-tops; deep canyons or clefts, from 1500 to 2000 feet sheer, like sudden rifts in the earth's surface; long secluded valleys, with forest-covered bottoms extending for many miles, and slopes of every conceivable gradient descending to a lower level of frightfully broken ground, joining the foot of the main range of Rocky Mountains at a distance of from 70 to 90 miles. There are also isolated patches of cotton-wood upon the sides of slopes, which afford excellent covert for deer and bears.

The actual width from margin to margin of the high land does not exceed 26 miles, although the length may be 100. It may readily be imagined that a month's shooting upon this area would be sufficient to scare the animals from the neighbourhood, more especially as the hunters are invariably on horseback, and traverse great distances each day.

When I was there we very seldom found bears upon the open, as they retired to the obscurity of the forests before break of day. Bob Stewart assured me that two seasons ago it was impossible to ride out in the early morning without seeing bears, but he counted up a long reckoning of seventy-two killed since the visit of Lord Lonsdale's party. This must have sensibly diminished the stock, and have afforded considerable experience to the survivors. Nevertheless upon several occasions bears exhibited themselves during broad daylight without being sought for.

We were tired of nothing but venison in every shape, and although the German cook, "little Henry," was a good fellow, he could not manage to change the menu without other provisions in the larder. I accordingly devoted myself one afternoon to shooting "sage-hens"; this is a species of grouse about the size of a domestic fowl, and, when young, there is nothing better. The old birds are not only tough, but they taste too strongly of sage, from subsisting upon the buds and young shoots of the wild plant. They were very numerous in certain localities, having much the same habits as the black game of North Britain, therefore we knew at once where to seek them.

Our camp was within a few feet of the little stream, just within the forest at the bottom of the valley; the dense mass of spruce firs extended for 8 or 10 miles along the slopes, only broken at intervals by gaps a few hundred yards wide, which divided the forest from top to base, and formed admirable places for ascending to the great plateau on the summit. This plateau extended for several miles, and was nearly level, the surface being liberally strewed with stones about 2 feet in length, but exceedingly flat, as though prepared for roofing slates; these had been turned over incessantly by the bears, in search for what Bob Stewart called "bugs"—the general and comprehensive American name for every insect.

We found a number of sage-hens upon this plateau, and I picked out the young ones with my rabbit rifle, as they ran upon the sage-covered ground. Texas Bill was soon loaded with game, and discarding the old birds that had been killed by mistake, we descended the grass-covered gap between the forests, and returned direct to camp. Little Henry had now a change of materials for our dinner.

It was nearly dusk, and I went into the small tent to have a hot bath after the day's work. I was just drying myself, after the operation of washing, when I heard an excited voice shout "Bears! bears!" It was useless for me to ask questions through the canvas, therefore I hurried on my clothes and ran out.

Texas Bill was gone. It appeared that two large bears had been seen as they came along the glen, and turned up the open slope, by which we had descended after shooting the sage-hens. My best horse had not been unsaddled, as the evening was chilly; therefore Texas Bill had immediately jumped into the saddle, and was off in full pursuit.

"What rifle did he take?" I inquired of little Henry. "He didn't take any rifle, but he's got his six-shooter, which is much better in his hands, as he knows it," was the reply.

There was very little light remaining, and with the long start which the bears obtained, I could not think that Bill would have the slightest chance of overhauling them before they reached the forest; this they would assuredly attempt, the instant they saw themselves pursued. If Bill could only get them upon the open plateau on the summit, he might be able to manage them, but with a gallop up a steep hill to commence with, in the late dusk of evening, the odds were decidedly against him.

It became dark, and we expected Bill's return every minute. Jem Bourne, my head man, who was always a grumbler, and exceedingly jealous, began to ventilate his feelings. "A pretty fool he's made of himself to go galloping after bears in a dark night, and nothing but a six-shooter! . . . A nice thing for our best horse to break his legs over those big rocks that nobody can see at night. . . . Well, he'll have to sleep out, and he'll find it pretty cold before the morning, I know. . . . What business he's got to take that horse without permission, beats me hollow!"

This sort of muttered growling was disturbed by two shots in quick succession, far up, above the summit of the forest. There could be no doubt that Bill had overhauled the bears.

By this time it was quite dark, and we drew our own conclusions from the two pistol shots, the unanimous decision being that Bill had fired in the hope of turning the bears when entering the forest; but what chance had he in the dark, and single-handed?

I did not take much interest in such a hopeless chase, but I was anxious about the horse, as the country was so rough that it would be most difficult to pick a way through holes and rocks, to say nothing of fallen trees, which, even during daylight, required consideration.

We piled immense pine-logs upon the fire, in addition to bundles of spruce branches; these made a blaze 20 feet high, and would form a beacon as a guide in the dark night.

I had taken the time by my watch when we heard the two shots upon the mountain top; twenty minutes had passed, and my lips were almost numbed by whistling with my fingers as a signal that could be heard during a calm night at a great distance. Suddenly this signal appeared to be answered by a shot, from a totally different direction from the first that we had heard; then, quickly, another shot; followed in irregular succession, until we had counted six. "His six-shooter's empty now, but he's got plenty of cartridges in his belt," exclaimed little Henry, the cook.

What was the object of these shots? He could not have followed the bears that distance in the dark, as his position was quite a mile from the spot where he had first fired; and he was now, as nearly as we could imagine, above a rocky cliff which bordered a grassy gap that would enable him to descend into our valley; he would then find his way parallel with the stream direct to our camp.

My men wished to fire some shots in response, but I declined to permit this disturbance of the neighbourhood, as it would have effectually driven all animals from the locality; we merely piled logs upon the fire, which could be seen from the heights at a great distance, and we waited in anxious expectation.

Nearly an hour passed away without any further sign. Bill could not have fired those six shots in succession to attract our attention, as it would have been a needless waste of ammunition: if he had expected a response to a signal, he would have fired a single shot, to be followed by another some minutes later. We now considered that he might have severely wounded the bear by the first two shots that we had heard, and that he had followed the beast up in some extraordinary manner, and at length discovered it.

We were about to give up all hope of his return, and knowing that he, as a smoker, was never without a supply of matches, we expected to see the glare of a distant fire, by which he would sit up throughout the night, when presently we heard the sound of whistling, and the clatter of a horse's feet among the stones of the brook, within 150 yards of our position.

In a couple of minutes Texas Bill appeared, leading the horse, which was covered with dry foam. In one hand he held a large bloody mass; this was the liver of a bear!

"Well done, Bill!" we all exclaimed, except the sulky Jem Bourne, who only muttered, "A pretty state you've brought that horse to; why, I shouldn't have known him."

The story was now told by the modest Bill, who did not imagine that he had done anything to excite admiration. This was his account of the hunt in the dark: "Well, you see, when the two bears were going up the open slope, down which you and I came, after shooting the sage-hens, all I could do was to gallop after them, to keep them from getting into the forest; when of course they would have been gone for ever. One of them did make a rush, and passed across me before I could stop him, and I didn't mind this, as I couldn't have managed two. I got in front of the other, and cracked my whip at him, and at last I got him well in the open on the big plateau, where we shot the sage-hens. He got savage now, and was determined to push by me and gain the forest; but I rode right at him, and seeing that I couldn't stop him, I fired my six-shooter to turn him, just as he made a dash at the horse. He made another rush at the horse, and I turned him with another shot, within a couple of paces' distance. This made him take off in a new direction, and he tried to cross the big plateau, intending, no doubt, to get to the forest a couple of miles away on the pointed hill. It was so dark that I could hardly see him, and my only chance was to ride round him, and work him till he should stand quiet enough to let me take a steady shot.

"He went on, sometimes here, sometimes there, and at last he changed his mind, and seeing that he couldn't get away from the horse across the open, he turned, and made for the 10 mile forest. It was as much as I could do to drive him, by shouting and cracking my whip whenever I headed him; if I had only once let him get out of sight, I should never have seen him again. The ground is full of stones, as you know, which bothered the horse in turning quickly;

but we went on, sometimes full gallop straight away, at other times dancing round and round, until at last the old bear got regularly tuckered-out, and he was so done he could hardly move. There he was, with his tongue hanging out of his mouth, standing, panting and blowing, and my horse wasn't much better, I can tell you. Well, I was drawn up as close to him as though I was going to strike him, and he was so completely done there wasn't any fight in him; my horse's flanks were heaving in such a way that I could hardly load the two chambers that I had fired. I was determined to have all my six shots ready before I began to fire, and it was just lucky that I did, for I'm blessed if I could kill him. There he stood, regularly exhausted-like, and he took shot after shot, and never seemed to notice, or to care for anything. At last I almost touched him, when I fired my sixth cartridge between his shoulders, and he dropped stone dead. That's all that happened, and I thought you wouldn't believe me if I came back without a proof; so I cut him open, and took out his liver to show you; and here it is."

Although this fine fellow thought nothing of his achievement, I considered it to be the most extraordinary feat of horsemanship that I had ever heard of, combined with wonderful determination. In the darkness of night, without a moon, to hunt single-handed, and to kill, a full-grown bear with a revolver, was in my experience an unprecedented triumph in shikar.

Early on the following morning I sent for the bear's skin. It proved to be a large silver-tipped, and a close examination exhibited the difficulties of the encounter during darkness.

Eight shots had been fired from the commencement, to the termination by the last fatal bullet; but, although Texas Bill was an excellent shot with his revolver, he had missed seven times, and the eighth was the only bullet that struck the bear. This had entered between the shoulders vertically, proving the correctness of his description, as he must have shot directly downwards. The bullet had passed through the centre of the heart, and had escaped near the brisket, having penetrated completely through this formidable animal.

Upon my return to England I immediately purchased a similar revolver of

Messrs. Colt and Co. — -the long frontier pistol, .450 bullet.

Although bears were scarce, we occasionally met them unexpectedly. As a rule, I took Jem Bourne and Texas Bill out shooting, the man Gaylord had to look after the twelve or thirteen animals, and little Henry, the German cook, was left in camp to assist my wife. Upon one of these rather dull days the camp was enlivened by the visit of three large bears. These creatures emerged from the neighbouring jungle, and commenced a search for food within 50 yards of the camp, only separated by a narrow streamlet of 10 feet in width. For about twenty minutes they were busily engaged in working up the ground like pigs, in search of roots or worms; in this manner they amused themselves harmlessly, until they suddenly observed that they were watched, after which they retreated to the forest.

My acquaintance Bob Stewart assured me that the bears had become so shy, that the only way to succeed was to "jump a bear." This term was explained as follows: you were to ride through forest, until you came across the fresh track of a bear; you were then to follow it up on foot, until you should arrive at the secluded spot where the bear slept during the daytime, in the recesses of the forest. It would of course jump out of its bed when disturbed, and this was termed "jumping a bear." Of course you incurred the chance of the animal's attack, when thus suddenly intruded upon at close quarters.

I agreed to start with Bob upon such an excursion; but I found that this kind of sport was more adapted for his light weight than my own, and that his moccasins were far superior to my boots, for running along the stems of fallen spruce trees at all kinds of angles, and for jumping from one prostrate trunk to another, in a squirrel-like fashion, more in harmony with a man of 9 stone than one of 15. We started together, Bob mounted upon his little mare, while I rode my best horse, "Buckskin," who was trained, like many of these useful animals, to stand alone, and graze, without moving away from his position for hours; should it be necessary to dismount, and leave him. The horses thus tutored are invaluable for shooting purposes, as it is frequently necessary to stalk an animal on foot; in which case, the bridle is simply arranged by drawing the reins over the head, and throwing them in his front, to fall upon the ground before his fore-feet. When thus managed, the horse will feed, but he will never move away from his position, and he will wait for hours for the return of his master.

We rode about four miles without seeing a living creature, except a badger. This animal squatted upon seeing the horses, and lay close to the ground like a hare in form, until we actually halted within 10 feet of its position. Bob immediately suggested that we should kill it, and secure its skin (his one idea appeared to be a longing to divest everything of its hide); but I would not halt, as the day was to be devoted to bears. We at length arrived at a portion of the forest where the young spruce had grown up from a space that had formerly been burnt; about 50 acres were densely covered with bright green foliage, forming a pleasing contrast to the sombre hue of the older forest. This was considered by my guide to be a likely retreat for bears; it was as thick as possible for trees to grow.

We accordingly dismounted, threw the reins over our horses' heads, and, taking the right direction of the wind, we entered the main forest, which was connected with the younger growth. It was easy to distinguish tracks, as the earth was covered with old half-rotten pine needles, which formed a soft surface, that would receive a deep impression. Nearly all the old trees were more or less barked by the horns of wapiti, showing that immense numbers must visit these woods at the season when the horns are nearly hard, and require rubbing, to clean them from the velvet. We had not strolled more than half a mile through the dark wood when Bob suddenly halted, and, like Robinson Crusoe, he appeared startled by the signs of a footstep deeply imprinted in the soil. It was uncommonly like a large and peculiarly broad human foot, but there was no doubt it was a most recent track of a bear, and the direction taken would lead towards the dense young spruce that we had already seen. We followed the track, until we at length arrived at the bright green thicket, in which we felt sure the bear must be lying down.

This was an exceedingly awkward place, and Bob assured me that if he were alone, he should decline to enter such a forest, as it was impossible to see a yard ahead, and a bear might spring upon you before you knew that it was near. As I had a double-barrelled powerful rifle, I of course went first, followed by Bob close behind. As noiselessly as possible, we pushed through the elastic branches, and very slowly followed the track, which was now more difficult to distinguish, owing to the close proximity of the young trees that overshadowed the surface of the ground.

In this manner we had advanced about a quarter of a mile, when a sudden rush was made exactly in my front, the young trees were roughly shaken, and I jumped forward immediately, to meet or to follow the animal, before I could determine what it really was. Something between a short roar and a grunt proclaimed it to be a bear, and I pushed on as fast as I could through the opposing branches; I could neither see nor hear anything.

Bob Stewart now joined me. "That's no good," he exclaimed, "you shouldn't run forward when you hear the rush of a bear, but jump on one side, as I did. Supposing that bear had come straight at you; why, he'd a been on the top of you before you could have got your rifle up. True, you've got a double-barrel, but that's not my way of shooting bears, although that's the way to JUMP A BEAR, which you've seen now, and you may jump a good many before you get a shot in this kind of stuff."

I could not induce Bob to take any further trouble in pursuit, as he assured me that it would be to no purpose: the bear when thus disturbed would go straight away, and might not halt for several miles.

This was a disappointment; we therefore sought our horses, which we found quietly grazing in the place that we expected. Remounting, we rode slowly through the great mass of spruce firs, which I had named the "10 mile forest."

There was very little underwood beyond a few young spruce here and there, and we could see from 80 to 100 yards in every direction. Presently we came across an enormous skull, which Bob immediately examined, and handed it to me, suggesting that I should preserve it as a specimen. He declared this to be the skull of a true grizzly; but some of the teeth were missing, and as I seldom collect anything that I have not myself shot or taken a part in shooting, I declined the head, although it was double the size of anything I had experienced.

The forest was peculiarly dark, and the earth was so soft from the decaying pine needles, that our horses made no noise, unless when occasionally their hoofs struck against the brittle branches of a fallen tree. We were thus riding, always keeping a bright look-out, when Bob (who was leading) suddenly sprang from his mare, and as quick as lightning fired at a black-tail buck, that was standing about 80 yards upon our right. His shot had no effect; the deer, which had not before observed us, started at the shot, and stood again, without moving more than three or four yards. Bob had reloaded his Sharp like magic, and he fired another shot, hitting it through the neck, as it was gazing directly towards us; it fell dead, without moving a foot.

We rode up to the buck; it was in beautiful condition, but the horns were in velvet, and were useless. I now watched with admiration the wonderful dexterity with which Bob, as a professional skin-hunter, divested this buck of its hide. It appeared to me that I could hardly take off my own clothes (if I were to commence with my greatcoat) quicker than he ripped off the skin from this beautiful beast. With very little delay, the hide was neatly folded up, and secured to the Mexican saddle by the long leathern thongs, which form portions of that excellent invention.

Bob remounted his mare, with the skin strapped behind the cantle, like a military valise; and we continued on our way. "That was a quick shot, Bob."—"Yes, 2 1/2 dollars, or 2 dollars at least I'll get for that skin; you see there's no game that pays us like the black-tail, and I never let one go if I can help it; they're easy to shoot, easy to skin, easy to dry, and easy to sell at a good price, and more than that, they're handy to pack upon a mule."

That little incident having passed, we again relapsed into silence, and rode slowly forward, with a wide-awake look-out on every side.

We had ridden about a mile, when the fresh tracks of bears that had crossed our route caused a sudden halt, and we immediately dismounted to examine them. They were of average size, and there could be no doubt, from the short stride of each pace, that they were retiring leisurely, after a night's ramble, to the beds in which they usually laid up. We led our horses to a small glade of good grass that was not far distant, and left them in the usual manner.

We now commenced tracking, which was simple enough, as the heavy footprints were distinct, and the bears had been travelling tolerably straight towards home. At length, after nearly a mile of this easy work, we arrived at a portion of the forest where some hurricane must in former years have levelled several hundred acres. The trees were lying about in confused heaps, piled in many places one upon the other, in the greatest confusion. None of them were absolutely rotten, but the branches were exceedingly brittle, and, if broken, they snapped like a pistol shot, making a noiseless advance most difficult. Through this chaos of fallen timber the young spruce had grown with extreme vigour, and I never experienced greater difficulty in making my way than in this tangled and obdurate mass of long trunks of gnarled trees, and branches lying at every angle, intergrown with the green boughs of younger spruce.

Bob Stewart wore moccasins, and being exceedingly light and active, he ran up each sloping treestem for 40 or 50 feet, then dropped nimbly to another fallen trunk below, bobbed under a mass of heavy timber, like masts in a shipbuilder's yard, supported as they had chanced to fall, and then dived underneath all sorts of obstructions. He was followed admiringly, but slowly, by myself, not provided with moccasins, but in high riding boots. If I had been a squirrel, I might perhaps have beaten Bob, but after several hundred yards of this horrible entanglement, which might have been peopled by all the bears in Wyoming, we arrived at a small grassy swamp in the bottom of a hollow, just beneath a great mass of perpendicular rock, about 70 or 80 feet in height. In the centre of this hollow was a pool of water, about 8 feet by 6. This had been disturbed so recently by some large animal, that the mud was still curling in dusky rings, showing that the bath had only just been vacated. We halted, and examined this attentively. The edges of the little pool were wet with the drip from the bear's shaggy coat, as it had left the water.

Bob whispered to me, "Look sharp, there are bears here, more than one I think, and if they've heard us, they'll be somewhere alongside this rock I reckon, or maybe up above." We crept along, and beneath the fallen timber; but it was so dark, owing to the great number of young spruce which had pushed their way upwards, that a dozen bear might have moved without our seeing one.

We now arrived at a small open space, about 20 feet square; this was a delightful change from the darkness and obstructions: The ground in this spot was a deep mass of pine needles, and in this soft material there were three or four round depressions, quite smooth, and about 18 inches deep; these were the beds of bears, where in undisturbed solitude they were in the habit of sleeping after their nocturnal rambles.

I was of opinion that we had disturbed our game, as several times we had accidentally broken a dead branch, with a loud report, when clambering through the abominable route. However, we crept forward round the base of the rock, and arrived in the darkest and thickest place that we had hitherto experienced.

At this moment we heard a sharp report, as a dead branch snapped immediately in our front. For an instant I saw a large black shadow apparently walking along the trunk of a fallen pine. I could not see the sight of my rifle in the deep gloom, but I fired, and was answered by a short growl and a momentary crash among the branches.

We ran forward with difficulty, but no bear was to be seen. We searched everywhere, but in vain. I came to the conclusion that the game was hardly worth the candle.

Through several hours we worked hard, but did not find another bear; and it was past five o'clock when we arrived at our camp, after a long day's work, in which we had certainly "jumped" two bears, but had not succeeded in bagging one.

Texas Bill came to hold my horse upon our arrival; he was looking rather shy, and ill-at-ease. "What's the matter, Bill? anything gone wrong?" I asked.

"Well," he replied, "I hope you won't blame me, as I don't think it right, but you know where you killed a wapiti a couple of days ago, and we found the next morning that the bears had been and buried it; and you said we'd better leave the place quiet for a day, and then you'd go early in the morning, and perhaps find the bears upon the spot? Well, after you were gone with Bob this morning, Jem Bourne proposed that we should go and have a look at the place, and sure enough when we got there we found a great big bear fast asleep, lying on the top of the buried wapiti, and her two half-grown cubs asleep with her. So Jem had your Martini-Henry with him, and he killed the mother stone dead, through the shoulder. Up gets one of the young ones, and hits his brother (or sister) such a whack in the eye with his paw that it just made me laugh, and then he cuffs him again over the head, just as though it was his fault that the mother was knocked over. Jem had reloaded, so he put a bullet through this young fellow; and then putting in another cartridge, he floored the third, and they were all dead in less than a minute. It's a fine rifle is that Martini-Henry, but I think you'll be displeased, as we had no business to go nigh the place; it ain't my fault, and I wouldn't have done it myself, you may be sure."

This was a glorious triumph for the jealous Jem Bourne, who was highly offended at my having adopted the advice, and sought the assistance of Bob Stewart, to "jump a bear." We had returned as failures, and he had killed three bears with my rifle, within my sanctuary, which I had specially arranged for a visit upon the follo-wing day. He declared "that nobody should stop him from killing bears, as his right was just as good as mine." This poaching upon my preserves was rather too much for my patience, therefore wit-hout any discussion or angry words I gave him a note to carry 42 miles' distance on the following morning to a friend of mine at the second ranche. "What horse shall I ride ?" asked the fellow sullenly. "The white mule," I replied. "When am I to come back ?"—"Not till I send for you," was the answer; and Jem Bourne ceased to be a member of our party.

This was an excellent example, as many of these people are exceedingly independent, and although he received high wages (120 dollars monthly, in addition to his food, and a horse to ride), he considered that he was quite the equal of his employer. Although my other men received only half these wages, they were more useful, and after this dismissal we were far more comfortable.

It was a strange study of the Far West in these outlandish and ut-terly uninhabited districts. When looking down from the summit of the mountains, facing north, we were positively certain that for more than 100 miles in a direct line there was not a human habitation, and the nearest point of embryo civilisation was the Government Park on the Yellowstone river, at least 150 miles distant. In our rear we were 80 miles from the abandoned station of Powder River, with only two ranches in the interval. It may be readily imagined that the laws of civilised communities were difficult to administer in such a wilderness.

The nearest railway station was "Rock Creek," about 240 miles, upon the Union Pacific, from whence we had originally started; that point is about 7000 feet above the sea-level. A curious contrivance, slung upon leather straps instead of springs, represents a coach, which, drawn by four horses, plies to Fort Fetterman, 90 miles distant. During this prairie journey the horses are only changed twice.

There are no dwellings to be seen throughout the undulating mass of wild grass; this possesses extraordinary properties for fattening cattle, and wild animals; but after a weary drive along a track worn by wheels and other traffic, and occasionally well defined by empty tins that had contained preserved provisions, a small speck is seen upon the horizon, which is declared to be the station for spare horses.

Upon arrival at this cheerless abode we entered a small loghouse, containing two rooms and a kitchen; but the cooking was conducted in the public room, an apartment about 13 feet square, with a useful kind of stove in one corner. The man who represented the establishment had of course observed the coach in the far distance, therefore he was not startled by the arrival of our party, which consisted of the Hon. Charles Ellis, Lady Baker, and myself. He had already begun to fry bacon in a huge frying-pan upon the little stove, and he had opened some large tins of preserved vegetables, in addition to another containing some kind of animal hardly to be distinguished. He had been successful that morning, having killed an antelope; therefore we had quite an entertainment in this log-hut, so far away from the great world.

The table was spread with a very dirty cloth, and our small party was immediately augmented by the arrival of the coachman (our driver), the man who looked after the horses, an outside passenger of questionable respectability, and our host, who had just cooked the bacon. It was an unexceptional fashion throughout the country to reduce all clothing to a minimum. Coats were unknown during the summer months (this was the middle of August); waistcoats were despised; and the costume of the period consisted of a flannel shirt, and a pair of trousers sustained by a belt in lieu of braces. Attached to this belt was the omnipresent six-shooter in its holster. I was the only person who possessed, or at all events exhibited, a coat; and I felt that peculiar and unhappy sensation of being over-dressed, which I feared might be mistaken for pride by our unsophisticated companions.

We were not a cheery party; on the contrary, everybody appeared to be so determined not to say the wrong thing, that they remained silent; the dullness of the meal was only broken at long intervals by such carefully expressed sentiments as "I'll trouble you to pass the salt, if you please," or "Will you kindly hand the bacon ?"

There was no vulgarity in this, and we were afterwards informed that these rough people, who, as a rule, season their conversation with the pepper of profanity, are painfully sensitive to the presence of a lady, before whom they are upon their P's and Q's of propriety; and, should an improper expression escape their lips in an unguarded moment, they would be in a state of deep depression from the keenest remorse, which might perhaps cause a sense of unhappiness for at least five minutes. They most sensibly refrained altogether from conversation in a lady's presence, to avoid the possibility of a "slip of the tongue."

If they could have left their perfume behind, together with the profanity, our table would have been sweeter; but the flannel shirts were seldom washed, to prevent shrinking, just as their owners seldom spoke, to avoid swearing; an overpowering smell of horses was emitted by the driver, and of stables by the ostler, while the proprietor exhaled the mixed but indescribable odours combined from his various duties, such as cooking, cleaning up, sleeping in his clothes, and never washing them.

The meal over, we again started. This stage was interesting, as we left the treeless expanse of prairie, and drove over highland through picturesque forests of spruce firs among rocks and canyons. About

20 miles of this scenery was passed; then we descended a long slope, and once more emerged upon the dreary, treeless prospect.

At the end of 35 miles another speck was seen, which eventually turned out to be a station similar to that at which we had halted in the morning. There were two pretty-looking and clean girls here; they had come to assist their brother, who "ran" the house. It was curious to observe the little evidences of civilisation which the presence of these girls had introduced. At first sight, among a rude community, I should have had strong misgivings concerning the security of young girls without a mother; but, on the contrary, I was assured that no man would ever presume to insult a respectable woman, and the girls were safer here than they would be at New York. It was a doughtful anomaly in a society which otherwise was exceedingly brutal, that a good woman possessed a civilising power which gained the respect of her rough surroundings, and, by an unpretentious charm, softened both speech and morals.

It was to be regretted that this benign influence could not have been extended to the vermin. When the lamp was extinguished, the bed was alive. I always marvelled at the phrase, "he took up his bed and walked," but if the bugs had been unanimous, they could have walked off with the bed without a miracle. Sleeping was impossible. I relighted the paraffin lamp, a retreat was evidently sounded, and the enemy retired. Presently an explosion took place — the lamp had gone wrong, and burst, fortunately without setting the place on fire. An advance was sounded, and the enemy came on, determined upon victory.

I never slept in one of those prairie stations again, but we preferred a camp sheet and good blankets on the sage-bush, with the sky for a ceiling.

On arrival at Fort Fetterman, 90 miles from Rock Creek station, the coach drew up at a loghouse of greater pretensions than those upon the prairie. I had letters of introduction from General McDowell (who was Commander-in-Chief of the Pacific Coast) to Colonel Gentry, who commanded Fort Fetterman, and Major Powell of the same station.

Not wishing to drive up to the door of his private house, we alighted at the log-hut which represented the inn. The room was horridly dirty, the floor was sanded, and there was a peculiar smell of bad drink, and an expression of depravity about the establishment.

The host was a tall man, attired as usual in a flannel shirt and trousers, with a belt and revolver. He had evidently observed an expression of disgust upon our faces, as he exclaimed, "Well, I guess we ain't fixed up for ladies; and p'r'aps it's as well that you came to-day instead of last night, if you ain't fond of shooting affairs. You were just looking at that table and thinking the tablecover was a bit dirty, weren't you? Well, last night Dick and Bill got to words over their cards, and before Dick could get out his six-shooter, young Bill was too quick and resolute, and he put two bullets through him just across this table, and he fell over it on his face, and never spoke a word. It's a good job too that Dick's got it at last."

This little incident was quite in harmony with the appearance of the den. I knew that letters had been previously forwarded from San Francisco to the Commandant, therefore I strolled towards his quarters, to leave my card and letter of introduction.

Fort Fetterman is not a fort, but merely an open station, with a frontier guard of one company of troops. I met Colonel Gentry, who was, very kindly, on his way towards the inn to meet us on arrival. Upon my inquiring respecting the fatal quarrel across the table, he informed me that he had held an inquest, and buried the man that morning.

The deceased was a notorious character, and he would assuredly have shot his younger antagonist, had he not been the quicker of the two in drawing his pistol.

This was a satisfactory termination to a dispute concerning cards, and there was a total absence of any false sentiment upon the part of the commonsense authority.

We were most hospitably entertained by Major and Mrs. Powell, to whose kind care we were committed by Colonel Gentry, who, being a bachelor, had no accommodation for ladies. It was very delightful, in the centre of a prairie wilderness, to meet with ladies, and to hear the rich contralto voice of Miss Powell, their daughter of eighteen, who promised to be a singer much above the average.

On the following morning we started for Powder River, 92 miles from Fort Fetterman; there was no public conveyance, as Powder River station had been abandoned since the Indians had been driven back, and confined to their reservation lands. We were bound by invitation to the cattle ranche of Mr. R. Frewen and his brother Mr. Moreton Frewen; these gentlemen had an establishment at Powder River, although their house was 22 miles distant upon the other side, in the centre of their ranche. They had very kindly sent a four-wheeled open carriage for us; one of those conveyances that are generally known as American waggons, with enormously high wheels of cobweb-like transparency. Jem Bourne had been sent as our conductor, having been engaged as my head man.

There was nothing but prairie throughout this uninteresting journey, enlivened now and then by a few antelopes.

Castle Frewen, as the superior log building was facetiously called by the Americans, was 212 miles from Rock Creek station, and we were well pleased upon arrival to accept their thoroughly appreciated hospitality. Their house had an upper floor, and a staircase rising from a hall, the walls of which were boarded, but were ornamented with heads and horns of a variety of wild animals; these were in excellent harmony with the style of the surroundings. Here we had the additional advantage of a kind and most charming hostess in Mrs. Moreton Frewen, in whose society it seemed impossible to believe that we were so remote from what the world calls civilisation. There was a private telephone, 22 miles in length, to the station at Powder River, and the springing of the alarm every quarter of an hour throughout the day was a sufficient proof of the attention necessary to conduct the affairs successfully at that distance from the place of business.

Our kind friends afforded us every possible assistance for the arrangements that were necessary, and we regarded with admiration the energy and perseverance they exhibited in working with their own hands, and in KNOWING HOW TO USE THEIR OWN HANDS, in the absence of such assistance as would be considered necessary in civilised countries.

There were about 8000 head of cattle upon the Frewens' ranche, all of which were in excellent condition. It was beyond my province to enter upon the question of successful ranching, but the Americans confided to me that the prairie grass, instead of benefiting by the pasturing of cattle, became exhausted, and that weeds usurped the place of the grass, which disappeared; therefore it would follow that a given area, that would support 10,000 head of cattle at the present time, would in a few years only support half that number. It might therefore be inferred that the process of deterioration would ultimately result in the loss of pasturage, and the necessary diminution in the herds.

From the Frewens' ranche, a ride of 25 miles along the course of the Powder river brought us to the last verge of civilisation; the utmost limit of the cattle ranches was owned by very nice young people, Mr. and
Mrs. Peters, Americans, and Mr. Alston, an English partner.

We had been hospitably received by these charming young settlers, whose rough log-house was in the last stage of completion, and I fear we must have caused them great personal inconvenience.

On the following morning we started for the wilds of the Big Horn, and crossing the Powder river, we at once commenced the steep ascent, for a steady pull of 4000 feet above the dell in which the house was situated. We left them, with the promise to pay them a few days' visit on our return.

It was then that we quickly discovered the peculiarities of our four attendants, whom I had expected to be examples of stern hardihood, that would represent the fabled reputation of the backwoodsman.

Although they were fine fellows in a certain way, they astonished me by their luxurious habits. In a country that abounded with game, I should have expected to exist upon the produce of the rifle, as I had done so frequently during many years' experience of rough life. A barrel of biscuits, a few pounds of bacon, and a good supply of coffee would have been sufficient for a crowned head who was fond of shooting, especially in a country where every kind of animal was fat. My men did not view this picture of happiness in the same light; they required coffee, sugar, an immense supply of bacon, an oven for baking bread, flour, baking-powder, preserved apples (dried), ditto peaches, ditto blackberries, together with the necessaries of pepper, salt, etc.

It was always my custom to drink a pint of cafe au lait and to eat some toast and butter at about 6 A.M. before starting for our day's work; after this I never thought of food throughout the day, until my return in the evening, which was generally at five or six o'clock.

My people were never ready in the morning, but were invariably squatted in front of the frying-pan, frizzling bacon, when I was prepared to start. Jem Bourne was a chronic grumbler because we hunted far away from camp, instead of returning at mid-day to luncheon. Excellent fresh bread was baked daily, and I insisted upon the people supplying themselves with sufficient food packed upon their saddles, if they were not hardy enough for a day's work after a good breakfast.

I observed that my friends Big Bill and Bob Stewart were also provided with a large supply of bacon, although they left the fattest animals rotting in the forest, simply because they hunted for the hides.

In the same manner I remarked the extreme fastidiousness of these otherwise hardy people in rejecting food which we should have considered delicious. I have seen them repeatedly throw away the sage-hens that I have shot; these were birds which we prized. On one occasion, as we were travelling when moving camp, I shot a jackass rabbit from the saddle, with my .577 rifle. It gave me considerable trouble to dismount and open this animal, which would have gained a prize for fat; having cleaned it most carefully, I stuffed the inside with grass, and attached it to the saddle. We never had an opportunity of eating this splendid specimen; on inquiring, the cook had thrown it away, "because at this season jackass rabbits fed upon sage shoots, and the flesh tasted of sage!

As we shall return to the Big Horn range when treating upon the habits of wapiti and other animals, I shall now refer to the Indian bears, and commence with the most spiteful of the species, Ursus labiatus.

CHAPTER XI

THE BEAR (continued)

The outline that I have already given of Ursus labiatus is sufficient to condemn its character; there are more accidents to natives of India and Ceylon from the attacks of this species than from any other animal; at the same time it is not carnivorous, therefore no excuse can be brought forward in extenuation. I have already observed that this variety of the bear family does not hybernate; it has a peculiar knack of concealment, as it is seldom met during the daytime, although perhaps very numerous in a certain locality. In places abounding with rocky hills, deep ravines, and thick bush, it may be readily imagined that bears obtain the requisite shelter without difficulty; but I have frequently visited their haunts, where no perceptible means of secreting themselves existed, nevertheless each night afforded fresh evidences of their industry in digging pits, when searching for white ants, within 150 yards of our camp. In these places we seldom found a bear, although driving the jungles daily with nearly two hundred beaters. This experience would denote that the bears travel long distances at night, to visit some favourite resort which produces the necessary food. The stomachs of all wild animals when shot should be immediately examined, as the contents will be a guide to the locality which they inhabit. I have killed elephants in Africa at least 50 miles distant from any cultivation, but their stomachs were filled with dhurra (Sorghum vulgare), thus proving that they had wandered great distances in search of a much-loved food that could not be obtained in their native forests. In the same manner all wild animals will travel extraordinary distances to obtain either water or food in countries where they are liable to be pursued. When the watchers who protect the crops are in sufficient force to drive the nocturnal intruders away with guns, the same animals will probably not reappear upon the following night, but they will visit some well-known spot in an opposite direction, and reappear forty-eight hours later upon the forbidden ground.

The elephants in that portion of Abyssinia which is traversed by the various affluents of the Nile, being much harassed by the sword-hunters of the Hamran Arabs, never drink in the same locality upon two nights consecutively; they drink in the Settite river perhaps on Monday, march 30 miles in retreat, and on the following night they will have wandered another 30 miles to the river Gash, in a totally opposite direction. They will then possibly return to the Settite, and after drinking, they will take a new departure, and march to the river Royan or to the Bahr Salaam.

A bear is a rapid traveller, and although sluggish in appearance when confined, it is extremely active; therefore outward signs of digging, although evidence of nocturnal visits, cannot be accepted as proofs of the bear's proximity.

I believe that leopards may be frequently crouching among the branches of trees, and remain unseen, while a person, unconscious of their presence, may pass beneath; but although the sloth bear is most active in ascending a tree, it would be difficult for it to remain unobserved, owing to its superior size and remarkable black colour. A very large old tree with a considerable cavernlike hole at the bottom should always be carefully examined, as bears are particularly fond of these impromptu dwellings. I knew a man who was thus surprised whilst cutting wood from a large tree, unconscious of the fact that a bear was concealed within the hollow trunk. The blows of the axe disturbed the occupant, which immediately bolted from the hollow, and seized the wood-cutter by the thigh. Fortunately the man had his axe, with which he at once belaboured the bear upon the head until it relinquished its hold. I saw the scars of the wound inflicted by the canine teeth; these were about 6 inches in length, extending from inside the thigh to the knee-joint. The man declared that if his axe had been heavier he could have killed the bear, but it happened to be exceedingly light, and had very little effect.

My shikari Kerim Bux, who was a very powerful man, had a serious encounter with a bear, which seized his master, and immediately turned upon him when he rushed unarmed to his assistance; the bear seized him by the leg, but in the wrestling match which ensued, Kerim came off victor, although badly bitten, as he threw the bear over a precipice, upon the edge of which the struggle had taken place. This man was head constable in the police, and bore a very high reputation.

The Ursus labiatus being one of the most vicious animals, I have seen it upon two occasions attack an elephant, one of which was quite unprovoked.

We had been driving jungle for sambur deer in the Balaghat district, and instead of posting myself upon a mucharn, or occupying any fixed position, I remained upon my elephant Hurri Ram. This was a tusker that had been lent to me by the Government upon two occasions, and he was so good-tempered, and active in making his way over bad ground in steep forests, that I determined to try him as a shooting elephant. I took my stand upon the open grass-land, which was beautifully undulating, and would have made a handsome park. Standing behind a bush we were partially concealed, and I waited in expectation that some animals might break covert in my direction. Presently I saw a dark object running through the low bushes upon the margin of the sal forest on my right, and a large bear emerged about 100 yards from my position. It stood upon the open for a few seconds, evidently taking a close scrutiny of the surroundings, prior to a run across the country, where no chance would be afforded for concealment. It suddenly espied the elephant, and, apparently without a moment's hesitation, it charged from the great distance of 100 yards at full speed directly upon the nervous Hurri Ram. I had not long to wait, but just as I pulled the trigger, when the bear was within 10 yards, the elephant whisked round and bolted down hill across the open, towards the portion of the jungle that was about 250 yards upon my left. Nothing would stop the runaway brute, but fortunately I had stationed a police constable at the very spot for which the elephant was making, and he, seeing the state of affairs, ran forward, shouting at the top of his voice and flourishing his rifle; this had the effect of turning the runaway, just as it was about to enter the forest, where we should in all probability have been smashed.

The bear had in the meantime gone across country, and although we hunted it for more than a mile, we never saw it again. This was a purely unprovoked attack, and it would have been interesting to have seen the result had the elephant not bolted. I imagine that the bear would have seized it by the leg, and afterwards would have attempted a retreat.

Upon another occasion, at a place called Soondah in the same district, I was upon Hurri Ram; I had been working through the high grass in the first-class reserves throughout the day, having killed a splendid stag sambur, when we were attracted by the peculiar short roar or moan made by a tigress calling either for her cub or for some male companion. This was in the sal forest, within a quarter of a mile of our position. It was a dangerous attempt, upon such an untrustworthy elephant as Hurri Ram, to look for a tiger in a thick sal jungle, as that species of tree grows in long straight trunks exceedingly close together, to an extent that would make it impossible for a large elephant to continue a direct course. Should the animal run away, the result would probably be fatal to the rider. We again heard the cry of the tiger repeated; this decided me to make the trial, and we entered the forest, carefully advancing, and scanning every direction.

The sal tree produces one of the most valuable woods in India for building purposes, and for railway sleepers. The bark is black, which gives the forest a sombre appearance, and the trees grow perfectly straight, generally to a height of 30 or 40 feet, before they divide into branches; it may be readily imagined that an elephant would find a difficulty in threading its way through the narrow passages formed by these mast-like growths. In addition to this difficulty, there were numerous clumps of the tough male bamboo, which nothing will break, and which is terribly dangerous should a runaway elephant attempt to penetrate it, as the hard wiry branches would lacerate a rider in a frightful manner. There were numerous ravines in this forest, and we kept along the margin, slowly and cautiously, peering at the same time into the depths, in the expectation of seeing the wandering tiger.

It was very perplexing; sometimes we heard the cry of the tiger in one direction, and upon reaching the spot, we heard it at a different place. I was determined not to give it up, and we worked for at least two hours, until we had thoroughly examined every ravine, and all the smaller nullahs that would have been likely hiding-places. "Past five o'clock," I exclaimed, upon looking at my watch. It was time to turn homewards, as it would be dark at six, and should we be benighted in the forest we should not find our way, neither would it be possible to ride an elephant, owing to the thick bamboo. We accordingly gave up our search for the tiger, and steered in a new direction towards the camp.

We had advanced for about half an hour through the gloomy forest, and were within about 3/4 of a mile in a direct line of the tents, when I observed a peculiarly dark shadow upon my right, about 35 yards distant, close to a dense mass of feathery bamboos. I stopped the elephant for an instant, and at the same moment the black mass moved away towards the thick cover of the foliage.

Guessing the position of the shoulder, I took a quick shot with the Paradox gun; the elephant, most fortunately, not having observed the animal.

The effect was most extraordinary; I never heard such a noise; there was a combination of roars and howls, as though a dozen tigers and lions were engaged in a Salvation Army chorus. Away went Hurri Ram, rendering it impossible for me to fire, as a large bear came straight at us, charging from the deep gloom of a bamboo clump, and growling, as it ran with the speed of a dog, direct at the elephant.

I thought we must be knocked to pieces; two or three smaller trees fortunately gave way before the terrified rush of Hurri Ram, but the power of the driving-hook was gone; although the mahout alternately drove the spike deep into his skull and hooked the sharp crook into the tender base of the ears, the elephant crashed along, threatening us with destruction, as he swept through bamboos, and appeared determined to run for miles.

I had been accustomed to feed this animal daily with all kinds of nice delicacies beloved by elephants, and at such times I always spoke to him in a peculiar phraseology. Although I was in the worst possible humour, and considerably anxious regarding our safety, when rushing through forest at 15 miles an hour, I addressed Hurri Ram in most endearing terms-"Poor old fellow, poor old Hurri Ram, where are the sugar-canes? where are the chupatties, poor old boy?" etc. etc. I believe thoroughly that the well-known tones of my voice restored his confidence far more than the torture of the driving-hook, and after a race of about 150 yards he stopped. "Now turn him round, give him the point sharp, and drive him straight for the bear." The mahout obeyed the order, and we soon approached the spot, where the roars and howls still continued. My men were up the trees; the shikari had thrown a mighty spear upon the ground, and had gone up the branches like a squirrel, as he did not see the fun of meeting the bear's charge.

Before we had time to examine the actual condition of affairs, the big bear suddenly dashed out again straight at the elephant, and once more in a disgraceful panic he took to flight, without the possibility, on my part, of taking a shot, when the bear thus daringly exposed itself. Again I had to comfort Hurri Ram, and by degrees we stopped his mad career, and once more returned to the scene of his discomfiture. There was a slight depression in an open hollow, where high grass in swampy ground intervened between two sections of the forest. As we advanced, the elephant being severely punished by the driving-hook and scolded by the mahout, the bear suddenly uprose from the high grass, and standing upon its hind legs, it faced us at about 40 yards' distance, affording a magnificent chance for a deadly shot. Away went Hurri Ram again, whisking round before I had a moment to fire; and after two successive chances of this kind, the bear escaped into the opposite jungle, and we searched for it in vain.

We now returned, and with some difficulty drove Hurri Ram to the scene of conflict. There was a bear lying dead. The howls and roars had ceased, and a few yards to the left of the dead bear was a large black mass: this was another bear, in the last gasp. Both had been knocked over by only one bullet from the Paradox.

Although I had only seen one bear, and that most indistinctly, it appeared that the bullet, being intensely hard, and propelled by 4 1/2 drams of powder, had gone completely through the shoulder of the original bear, and then struck an unseen companion, who must have been some yards distant upon lower ground beyond. The bullet had broken the shoulder of this unlucky friend, and was sticking in its lungs, having carried a bundle of coarse black hair from bear No. 1 and deposited it upon its course in bear No. 2.

Although these were full-grown bears, there can be little doubt that the bear that had so determinedly attacked the elephant was the mother, infuriated by the roars and howls of her dying offspring. The penetration of the Paradox bullet was highly satisfactory, but I was terribly disgusted with Hurri Ram, whose misconduct had caused the loss of bear No. 3, which would most certainly have been included in the list of killed had I had the chance of only one second's quiet.

My men were not in the least ashamed when they descended from the trees, as they considered that the better part of valour was discretion. The large spear had been manufactured expressly for this kind of emergency, by a celebrated native cutler, Bhoput of Nagpur. It is always advisable that some powerful and plucky shikari should carry such a weapon for approaching any wounded animal, as accidents generally occur from carelessness, when the animal is supposed to be lying helpless, at the point of death. Such a spear should be 2 feet long, with a blade 3 inches wide, and extremely sharp. There should be a short cross-bar about 22 inches from the point, to prevent the spear from running completely through an animal, which could then writhe up the handle, and attack. The socket should be large and long, to admit a very thick male bamboo, as the mistake is too frequently made that the spear is strong, but the handle is too weak. It is very important that a trustworthy attendant should be thus armed, as a dying animal can then be approached with comparative impunity.

The risks that are run in following wounded animals are far greater than the prime attack. Should an animal charge without being wounded, it may generally be turned by a steady shot, if not absolutely killed; but when badly hurt, the onset of a beast is spasmodic, and nothing but death will paralyse the spring. I could mention numerous cases where lamentable disasters have occurred simply through thoughtlessness on the part of the hunter, who has been sacrificed in consequence of his neglect. One of the saddest catastrophes was the death of the late Lord Edward St. Maur, son of the Duke of Somerset, who died from the effects of amputation necessitated by the mangled state of his knee from the attack of a bear some years ago in India. This unfortunate young sportsman was shooting alone, and having wounded a bear, he followed up the animal for about a mile. When discovered it immediately charged him, and although again seriously wounded by his shot, the bear seized him by the knee, pulled him to the ground, and in the struggle that ensued he was seriously mauled. The bear was driven away by his attendants, and he was conveyed to camp. There was no blame in this instance attached to himself, or to any other person. In a most courageous manner he defended himself against the bear with his hunting-knife, and the body of the animal was recovered after some days by his shikari; but this promising young nobleman was cut off in the early days of his career, and was probably sacrificed through a want of surgical experience on the part of the native operator. I remember an instance of carelessness, which might have had a disastrous result, many years ago, when I was hunting in Ceylon. My brother, the late General Valentine Baker, was riding with me through the jungles in the district called "The Park." I had been caught by a rogue elephant a few days before, and my right thigh was so damaged that I could only walk a few yards with difficulty. Suddenly the man who walked before my horse ran back, and shouted "Wallahah, Wallahah" (Bears, Bears), and we caught sight of some large black object rushing through the jungle, close to our horses' heads. Valentine Baker jumped nimbly off, and I heard a shot almost immediately; my wounded leg was perfectly numbed, and I had no feeling in my foot; therefore, as it touched the ground without sensation, I fell over on my back. Gathering myself together, I managed to run in chase, and I shortly found myself close to the retreating heels of two bears that were trotting through the dense underwood. One of these brutes, feeling that it was pursued, turned quickly round, and immediately jumped upon the muzzle of my gun, which I fired into its stomach and rolled it over. I now heard my brother shouting my name at only a few yards' distance; running towards him, as I feared some accident, I found a large bear half lying and half sitting upon the ground, growling and biting at the hard-wood loading-rod which V. Baker had thrust into a bullet wound behind its shoulder; he seemed surprised that the bear would not die at once. This was exceedingly dangerous, as the animal might have recovered sufficient strength to have directed an attack at an unguarded moment. Having a heavy hunting-knife of 3 lbs. weight, I gave it a blow across the skull, which cleft it to the brain and terminated its struggles. This was exactly the occasion upon which an accident might have occurred, and when a spear would have been of use.

I cannot understand why persons who reside in India neglect the assistance of dogs for the various kinds of hunting. Bull terriers would be invaluable for tracking up a wounded tiger or bear, and the latter might be hunted by such dogs even without being wounded. At any rate, well-trained dogs would be of immense assistance, but I have never seen them used. During the cool season of Central and Northern India the climate is most favourable, and the dogs could work during the hottest hours of the day without undue fatigue. Mr. Sanderson set the example some years ago, and had some interesting hunts; he describes the Ursus labiatus as rendered powerless, in spite of its great strength and activity, as one bull terrier invariably seized it by the nose; this is the most sensitive part, and easy to hold, as it is long, and connected with a projecting upper lip, which is almost prehensile in this variety. His experience proved that three dogs were sufficient to hold any bear, as the claws, although dangerous to the tender skin of a man, were too blunt to tear the tough but yielding hide of the dog.

There are two other varieties of bears in the continent of India, the black (Ursus Thibetanus) and the brown, both of which are confined to Cashmere and the Himalayah range. I have had no personal experience of these animals, therefore I do not presume to offer myself as an authority; but from the accounts I have received from those who have hunted them successfully, they are much the same in their habits as the average of their species.

The dangerous character of bears, in like manner with all other animals, was accredited at a time when breechloaders and high velocities were unknown, but with a '577 rifle and 6 drams of powder, or a No. 12 spherical and 7 drams of powder, I cannot conceive the possibility of escape for any bear or other creature below the standard of a buffalo, if the hunter is a cool and steady shot. The conditions of this theory will include a solid bullet, not a hollow projectile dignified by the term "Express."

I will conclude this notice of the bear with an example of the failure of the hollow bullet, '577 Express, fired by a native gentleman,

Zahur al Islam, when shooting with me in the reserves of Singrampur in the Central Provinces last winter.

We were driving for any kind of animals that the jungle might produce, and, being on foot, we constructed the usual little hiding-place by cutting half through a sapling about 3 feet from the root, and bearing down upon the young tree so as to form a horizontal rail in front of our seat; a similar cut at the back of another sapling about 3 inches thick, facing the stem already laid, and that was also pressed down to interlace with the branches of the prostrate tree. This makes a screen which can be rendered still more opaque by the addition of a few green boughs.

The grass was parched to a bright straw colour, and was about 4 feet high. As the beaters approached, a bear rushed forward and passed within 15 paces of Zahur. He fired; the bear emitted a short growl and passed on.

I assisted in tracking this animal by the blood upon the grass. Zahur described the shot he had taken as oblique; as the bear had passed him, therefore the bullet must have struck either the hindquarters full, or the thigh.

We found a teak tree about 14 inches in diameter covered with small pieces of flesh resembling sausage-meat, for a height of 6 feet from the ground. The yellow grass at the foot of this tree was covered with blood, and many minute fragments of flesh adhered to the leaves. Searching the place carefully, we picked up two pieces of bone covered with blood; these were very thick and strong, the larger fragment being 2 1/2 inches in length and 1 inch in width, evidently pieces belonging to the upper portion of the thigh.

After tracking the wounded bear for about 200 yards through the high grass and jungle, we came to a tolerably deep nullah, where we expected to find the animal lying down. Instead of this, we discovered another large piece of fractured thigh bone, which proved that the hollow Express bullet, although '577, had broken up upon striking the bone, instead of penetrating throughout the body. The muscles of the thigh and the bone had been shattered to atoms, and the flesh so completely exploded that it had flown in all directions, dispersed in the smallest fragments; nevertheless this bear had gone right away, and was never more seen, although we expended more than an hour in its search, both with men and elephants.

There could not be a more cruel example of the effect of a hollow projectile when striking a bone. If that had been a solid bullet, it would have raked the animal fore and aft, and would have rolled it over on the spot.